SYSTEMS SIMULATION IN AGRICULTURE

11·50

G. Hughes

SYSTEMS SIMULATION IN AGRICULTURE

J. B. DENT, Ph.D.

Professor of Farm Management and Rural Valuation
Lincoln College, Canterbury, New Zealand

and

M. J. BLACKIE, Ph.D.

Senior Lecturer in Agricultural Economics,
School of Agriculture, University of South Pacific, Western Samoa

with assistance from

S. R. Harrison, Ph.D.

University of Queensland, Australia

APPLIED SCIENCE PUBLISHERS LTD
LONDON

APPLIED SCIENCE PUBLISHERS LTD
RIPPLE ROAD, BARKING, ESSEX, ENGLAND

British Library Cataloguing in Publication Data

Dent, John Barry
Systems simulation in agriculture.
1. Agriculture—Simulation methods
I. Title II. Blackie, M J III. Harrison, S R
630'.1'84 S494.5.S/

ISBN 0-85334-827-8

WITH 8 TABLES AND 48 ILLUSTRATIONS

Printed in Great Britain by Galliard (Printers) Ltd, Great Yarmouth

Contents

Preface

A fast-growing interest in the concepts and application of systems research has spawned a wide and general literature over the past decade. Most disciplinary areas have been touched, but commerce, engineering and military studies have, perhaps, been best served with outstanding texts. No provision has so far been made for a general book at introductory level of direct relevance to agricultural science, technology and management. General reviews are, of course, valuable to the agricultural-systems researcher but agricultural systems, with important biological components interacting with equally vital social and economic elements, embody particular characteristics which influence the approach to their study. This book is written in the belief that the concepts as well as the technology of the systems approach have a basic role in the rational advancement of the agricultural discipline and in the improvement of efficiency in agricultural research and practice. A basic and introductory text is an essential pre-requisite to this role being realised.

A reiteration of basic concepts is expressed in the introductory chapter while in the final chapter particular attention is given to the general problems of integrating systems concepts in research, extension and practice. The dialogue of these chapters is necessarily brief and in some respects speculative but it is supported by appropriate bibliography.

The main body of the text is concerned with the methodology of systems research; the conception, construction, implementation, validation and exploitation of computer-based simulation models of agricultural systems. The book has been structured for ease of reference for the agricultural researcher who has had little opportunity to become involved in systems studies and who would like to become familiar and operational in this area. It is also hoped that the experienced modeller will find something of value in the book; perhaps a refreshment of ideas, perhaps inspiration to correct entrenched faults, perhaps some additional view of familiar procedures or perhaps, through disagreement with our view, he may gain further insight into general methodology or application.

Obviously, we anticipate that readers will have a general appreciation of agriculture as well as a firm understanding of their own speciality and will have a basic grounding in statistical method. However, no previous experience in model-building is assumed though those with some association with computers and a computer language will find progress through the book simpler and quicker than those without this experience. In general, it would be an advantage to have completed an introductory course in computer methods before detailed study of this book. In the belief that learning is more effective by involvement and practice, the text is reinforced by reference to two detailed examples and it is hoped that those without previous association with computers will find these particularly valuable. The two different examples are described and defined in appendices to early chapters and are gradually woven into the text to assist in the explanation of various procedures.

The text has been extended from an original draft by the authors' association with Dr Steve Harrison whose recent work at the University of Queensland on experimental methods with computer models has been built into Chapter 6. We are most grateful for his involvement with this chapter as well as for many helpful comments on early drafts of the book. Data for one of the detailed examples used throughout the text were kindly provided by Dr P. S. Teng and Dr Ron Close of Lincoln College. Dr Tony Bywater, University of California, Dr Peter Chudleigh, Lincoln College, Dr Gary Fick, Cornell University, Dr Steve Filan, University of New South Wales and Dr P. S. Teng all made valuable contributions to earlier drafts. Many other colleagues and associates helped in a variety of ways to formulate and clarify the ideas expressed in this book; residual inadequacies and omissions remain our responsibility.

We would like to thank Ruth Frampton for transforming our incredibly rough pencil sketches into appropriate illustrations and Marion Mischler for patience and customary skill in typing the drafts.

J. B. DENT and M. J. BLACKIE
Lincoln College

1

Principles of Model-Building

For most people, discussion concerning a model provokes thoughts of some kind of physical representation of a real object—usually in miniature form. Such models will look like and often function in a similar (though often markedly simplified) way to the real object. Such physical models (classified as 'iconic' by Churchman, 1971) embrace children's toys, tailor's dummies and mock-ups of buildings and structures to be later constructed in real form. Representation of town planning developments or of the prospective layout of controls in a nuclear submarine are examples of useful iconic models and illustrate the point that models can be constructed of objects or situations not yet in existence in real form. It is not only miniature or life-size iconic models which prove useful; chemists, for example, find it expedient to construct greatly magnified physical representations of the structure of complex molecules. There is, however, an extremely useful class of model which bears little or no physical resemblance to the system it is intended to represent and it is this type of model which will be our major concern in this book. Churchman (1971) classifies such models as 'symbolic'. Symbolic models are abstract in form and are perhaps more difficult to comprehend than iconic models. The abstraction of the model frees it from the limits of physical form, thus rendering this type of model considerable flexibility in both the mode of its construction and the manner of its use.

Chemists will often represent molecular structure in symbolic form as part of the statement of a chemical reaction such as, for instance, the example below.

$$2H_2O + 2Cl_2 = 4HCl + O_2$$

Here we have an example of a model whose symbolic manifestation bears no physical resemblance to the real form; it is nevertheless a direct representation of the real chemical reaction and hence should be considered as a model of the reaction.

A fundamental principle of model-building, therefore, is that the type of model to be constructed depends on the use to be made of it: the model should represent those facets of the real system relevant to the model-uses. So iconic models of molecular structure, while valuable to the physical chemist, are translated into symbolic form by the industrial chemist in the study of chemical reactions.

Model-building is not an exact science: indeed, Mihram (1972) defines it as the 'art of mimicry'. Given that the function of a model is to mimic the behaviour of a real object or situation then the determination of what format the model should take and what degree of detail should be represented remains a matter of judgement on the part of the model-builder. In this text, we will be considering the issues faced by the model-builder in all aspects of his work. Guidelines to assist the modeller in making the necessary value judgements will be presented together with examples drawn from actual modelling experience.

Symbolic models are diverse in form, and classification has been attempted by a number of authorities (see, for example, Anderson, 1974; Mihram, 1972; Shannon, 1975; and Throsby, 1973 in the reference section at the end of this chapter). This text, however, is confined to a single form of symbolic model which we will call a computer-based simulation model. This model-type has particular strength in mimicking complex situations characterised by uncertainty and change over time. (Model-builders refer to such situations as being complex, stochastic and dynamic.) In the last decade, computer-based simulation models have gained acceptance in most branches of learning and have found application in fields as separate as defence-system development, corporate management and ecology. Many of the concepts and approaches set out in this text have been developed in other disciplines and have been adapted for application in an agricultural context. Of course, many other disciplines are beset with the same basic need to consider change in uncertain circumstances; to this extent the procedures developed in this book may, in principle, find application in diverse areas outside agriculture. Reference will be made to studies seemingly unconnected to agriculture but which have value in illustrating particular aspects of simulation practice. For the most part, however, examples will be drawn from two models—a detailed biological model of barley-leaf-rust disease and a more pragmatic management-orientated model of a pig-production unit. Between them, these two models include many features of interest to biological and agricultural model-builders. These two models are presented in detail in the appendices at the end of this and subsequent chapters. The reader may find it useful to refer to these appendices from time to time in reading through this text.

Why the Computer?

Simulation models may be divided into two classes, predictive simulation models and mechanistic simulation models: the first is concerned with providing forecasts (predictions) relating to alternative options, the second involved with assisting understanding of the situation under study, perhaps with a long-term view of controlling it more effectively. Both predictive and mechanistic simulation models can take a number of forms, one of which is computer-based. The computer then is not essential in simulation modelling. Indeed in agriculture at present, it would be the exception rather than the rule.

Consider the barley-leaf-rust model described at the end of this chapter. Once the farmer has planted a crop of barley, that crop is susceptible to leaf rust during certain parts of the growing season. This disease can severely depress yields depending on the stage of crop growth and weather conditions when it attacks. Present New Zealand farming practice is to spray the crop with fungicide at regular calendar intervals during that part of the season when disease outbreaks are most likely to occur. This practice is based on a very simple 'mental' simulation model which assumes

1. disease will always occur in the crop;
2. yield reduction and hence financial losses from the disease will be heavy in the absence of fungicide application.

Where fungicide and application costs are low, this simple mental model may be quite adequate. However, as the margin between cost of spraying and the potential economic yield loss caused by disease narrows, so it becomes important to construct a more sophisticated model to direct decisions about fungicide application. Such a model would include a disease-loss relationship which is dependent upon past and future weather patterns and crop growth stage. It would then be possible to use this relationship to examine the dynamics of the disease epidemic on a specific crop. The complexities involved in such a model render it capable of expression only by implementation on a computer.

The computer is ideally structured both to hold and recall vast amounts of information and to follow faithfully the changing state of many interacting variables over time. A simulation model that exploits these characteristics of the computer can have operational advantages over the mental simulation model. Provided the data base is adequate and the computer-based simulation model is appropriately constructed and programmed, the assessment of alternative decisions will be more comprehensive. The operational value of such a model will depend on factors such as the cost of its construction, the ease with which the decision maker can have access to it and the added value from more comprehensive assessment of options. The computer is an extension of human mental capacity (not a replacement for it) and the computer-based simulation model permits a more formal consideration of the information pertaining to a decision. This imposes on the modeller a responsibility to ensure that his data base is compatible with the complexity of the system under study, that his use of the data is valid, that his construction of the model is not biased and is related to current knowledge and that his computer programming is accurate. Many of these demands remain in the realm of subjective judgement by the modeller. The form of this book may now be seen in rather more detail since it presents

1. concepts to assist these judgements;
2. methods to ensure model-construction proceeds logically and efficiently;
3. procedures for handling data for modelling;
4. approaches for testing the validity of models in relation to their required function; and
5. a philosophy for the application of computer-based simulation models to agriculture.

A Definition of System

The word system is commonly used in a general sense and usually the meaning inferred is of a complex set of related components within an autonomous framework. We have all used terms such as the 'telecommunications system' or the 'transport system'; in agriculture, we are familiar with the meaning of 'harvesting system' or 'metabolic system'. The term defines, in a general way, the limits of autonomy and it implies that within these limits there is an unprescribed complex organisation. Usually it is assumed that the organisation is established or exists in order to carry out or maintain some function(s) (even if this is simply to maintain an existing equilibrium between the components of the system).

The striking implication is that the complex interrelationships between components precludes legitimate study of sectors of the system in isolation. Because the interrelationships are so important, the whole system is more complex, more comprehensive, than the sum of the individual components. Protagonists of the system approach agree that this is a fundamental fact applying to all systems and is the unifying theme in systems theory. Any defined system will have its own specific characteristics but all systems will conform to the following general features:

1. A system is fully defined both by a set of identifiable entities (or components) and interconnections between them and by the limits to their organisational autonomy.
2. A system is a hierarchical structure comprising a number of subsystems each capable of autonomous definition; in turn subsystems similarly embody the next layer of detail in autonomous sub-subsystems. The point of entry into the hierarchy in any systems study is related to the objectives for which the system is being studied. The number of layers of the hierarchy included in any study will depend on the judgement of the researcher but certain rules of thumb have been established. These will emerge in later discussion. (See also de Wit, 1970).
3. The most important characteristics of systems emerge over time so that the understanding of systems requires explicit consideration of time and rates of change.
4. Systems are sensitive to the environment in which they exist. This environment is usually unpredictable and certainly variable.

Systems research is therefore the study of systems: it involves, first, an analysis of the components and relationships of a system and, secondly, a synthesis phase. This latter phase might involve either the development of new systems or the more efficient use of the original system. Analysis and synthesis are usually not distinct in time; rather a cycling between the two develops (Wright, 1971). Systems research generally involves the construction of a computer-based simulation model. Obviously, such a stage will follow considerable analysis of the system, but after a certain point the model itself can begin to guide analysis so that experimental resources are more effectively used. To be effective the model not only must be technically sound but also must faithfully present the four general characteristics outlined above. The model is an integral part of systems research, acting as

1. a medium by which experimental studies can be guided;
2. a method whereby the results of such work are accumulated (and assessed);
3. a platform to guide the development of new systems or to control (assist decision making) in existing systems.

One basic component of the structure of any system is its boundary. A system's boundary is a contrived component designed to assist the understanding of the system's function; in reality, of course, no such boundary exists on any system. However, without a precisely defined boundary, the performance in terms of output from the system in relation to inputs cannot be determined by way of a model. The placing of the boundary is of considerable importance in modelling the system since it determines exactly which subsystems must be explicitly represented within the model-structure. Across the boundary there is assumed to be no interaction. Hence, in the model it is assumed that input to the system across the boundary provides the total environment in which the system (in model-form) must operate. The data requirements to provide for this environment are quite different from those required to construct the detailed subsystems. The model-environment will be composed of a number of elements which will be termed *exogenous* variables or '*driving*' variables. So far as the model is concerned these exogenous variables represent the only 'contact' with the circumstances outside the defined system. Placing of the boundary in modelling work will depend primarily on the function the model has to fulfil but will be affected by data availability; where, for example, paucity of data means that a subsystem cannot be represented adequately within the boundary, it may be convenient to contract the boundary so that this element of the model is removed from the interacting complex of the system and reduced to the form of an exogenous variable.

Consider again the barley-leaf-rust model. A plant-disease epidemic is a complex parasite–host system where growth of parasite and host, and their interaction, are affected by a large number of factors. The system in this example may be defined as the growing crop and is represented pictorially in Fig. 1.1. The conceptual boundary is drawn around the crop and the system consists of two major subsystems:

1. the fungus *Puccinia hordei* Otth;
2. the crop *Hordeum vulgare* L. (barley).

Since the disease is foliar, the crop subsystem in the model is represented by plant leaf area and progress of the disease is calculated from the percentage of leaf area infected by disease. In the fungus subsystem, the organism assumes different forms during its life-cycle, creating a multicomponent system in which the condition of each component is measured by the number of fungus bodies in each life-cycle form. Exogenous variables are inputs of weather, fungicide and external inoculum and these generate outputs of spores and crop yield across the boundary.

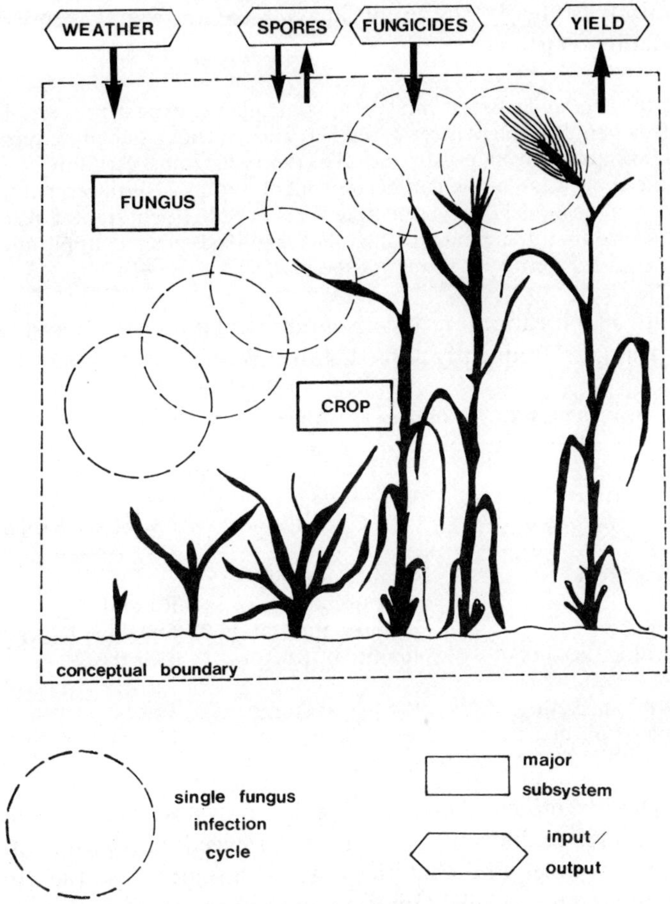

Fig. 1.1. *The leaf rust–barley crop agroecosystem.* (*After Teng, Blackie & Close,* 1978.)

Exogenous variables may usefully be classified as either those which are controllable or those which are uncontrollable by management. The importance of this classification is that it leads directly on to the way in which a particular exogenous variable is represented. Inputs to the system through the boundary that are fully under the control of management are related to policy decisions. Such controllable exogenous variables represent policies the outcome of which the model may be used to assess. Other controllable exogenous factors which are policies that the model will not be used to study explicitly can be considered fixed and simply represented in the model by constants.

In the barley-leaf-rust model, a controllable exogenous variable is fungicide application whereas weather and outside inoculum represent uncontrollable exogenous variables. The crop will almost certainly be grown with the application of such other controllable inputs as fertiliser, but for the purpose of the model this type of input is assumed to be constant. Effectively, this means that the model-builder has assumed normal application of fertilisers to the crop subsystem of the model.

Uncontrollable exogenous variables represent inputs to the system which are uncertain and frequently unpredictable. We will be discussing how such variables may be modelled in Chapter 4. For now it is sufficient to note that a greater degree of difficulty is usually experienced than with controllable exogenous variables and that the data requirement for the stochastic specification of uncontrollable exogenous variables can be extensive. Some examples of controllable and uncontrollable exogenous variables are given in Table 1.1.

Table 1.1. Examples of various exogenous variables

System definition	*Exogenous variable*	
	Controllable	*Uncontrollable*
Grazing livestock	Number of stock per hectare	Rainfall
Crop growth	Fertiliser application	Sunshine hours
Mechanical harvesting of grain	Number of tractors	Wet days
Commodity marketing	Number of stores	Prices
Rumen metabolism	Feed offered	Dry-matter consumption

The outputs from the system are dependent on the stream of exogenous variables imposed upon the system and the structure and organisation of the system itself. In some managed systems the value of the outputs will determine future direction of the controllable inputs. This structure implies an information feedback to the manager in time for him to make adjustments to his management policy. Information systems which have this *cybernetic* (feedback) characteristic are discussed in detail in Chapter 7. Here we should merely note that the value of outputs at one point in time will, in the presence of an information feedback, influence the

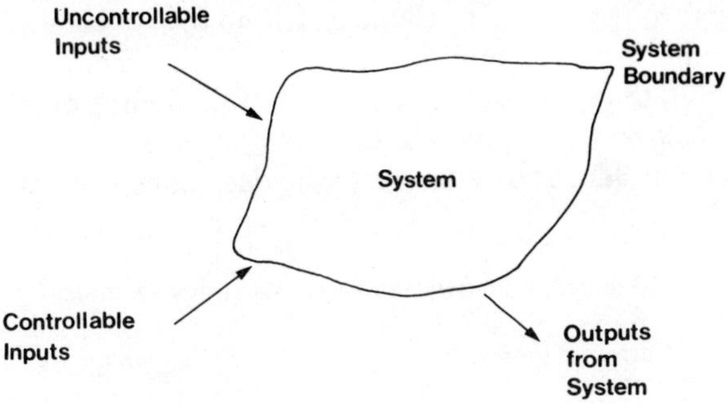

Fig. 1.2. *The concept of a system.*

value of outputs at some later date. The general system may then be represented in schematic fashion as in Fig. 1.2.

To illustrate briefly this concept of feedback, look now at the pig-management model described at the end of this chapter. This model is intended to provide management information on the results of particular strategies adopted by the farmer. The model accepts inputs of feed policies, herd structure, fertility, etc., and generates outputs of production and cash flow from the unit. Consider the case of a manager wishing to examine the economic implications of a different feeding policy for his fattening pigs. The new feed will be cheaper than his present one but will depress the growth rate of the fattening pigs. If his fattening house is already used to capacity, the new feed policy will result in overcrowding in the fattening house. This uncontrolled cybernetic loop will result in increased mortality, injury from stress diseases, such as tail biting, and further depressed growth rate. However, there is a controlled feedback option which is to reduce the input of pigs into the unit to compensate for the reduced outflow of pigs from the unit. The model allows the farmer to examine the economic implications of these and other management options.

The barley-leaf-rust model also provides an example of feedback loops. Assume a crop is infected by an initial dose of inoculum giving rise to a focus of infection. Pustules erupt at this infection site and produce infective spores. Given suitable conditions of weather and crop susceptibility to infection, these spores will germinate in the crop foliage and produce yet more infective pustules. This *positive* feedback gives rise to a disease epidemic. At some stage during the epidemic, a *negative* feedback loop comes into play bringing the epidemic back under control. This negative feedback may be direct management action in fungicide application, or simply occur because so much of the foliage is infected that there are few or no sites left for further infection.

Under the influence of the exogenous variables the state of the system will be changing through time. Key elements within the system will be changing their values and it is from monitoring these values (either in the real system or a model of it) that an understanding of the functioning of the system may be gained. Such key elements are normally termed *status* variables. The condition of a system at any time may be determined by establishing the

Table 1.2. Examples of various status variables

System definition	*Status variables*
Grazing livestock	Liveweight stock
Crop growth	Leaf area index
Mechanical harvesting of grain	Grain harvested per day
Commodity marketing	Amount of commodity in stores
Rumen metabolism	Volatile fatty acid formation

values of the status variables. Examples of some status variables are given in Table 1.2.

In the barley-leaf-rust model, an important status variable is the number of spore-producing bodies present in the crop. In the pig model, status variables include the number of sows in the various stages of lactation and pregnancy.

System Simulation

Strictly speaking, the application of mathematical programming models (such as linear programming) is encompassed by simulation. After all a model, adjudged suitable by the designer, is constructed to represent a system and is used to examine changes in that system. However, a problem of suitability of model-type is involved and most systems researchers would suggest that the following points be carefully weighed if mathematical programming models are being considered for a simulation study:

1. The fixed framework of mathematical programming methods imposes a rigidity of model-structure.
2. Incorporation of detailed interactive data (e.g. biological relationships) is often cumbersome within mathematical programming methods.
3. Mathematical programming models are solved in relation to a

specified criterion. Multicriteria mathematical models are cumbersome and computer-time absorbing unless these criteria can be expressed as a linear combination.

4. The stochastic and dynamic elements are not simply (or cheaply, in terms of computer usage) incorporated into mathematical programming models.

The many discussions and alternative points of view relating to these points may be read elsewhere (see the further reading section at the end of this chapter) and each model-builder must decide if a mathematical programming framework is appropriate for his purpose.

What we are calling computer-based simulation models are free-form models which are not encumbered with any of the problems of the mathematical programming model. These models are 'run' rather than 'solved' which implies that use of the model involves learning by exploration of the system vicariously via the model. Ackoff (1973) has coined the phrase 'decision-support' and it is an apt one in relation to computer-based simulation models; it implies that the model provides objective information in guiding decision making or in extending understanding of the system. In this regard all such models are 'input–output' in nature, in that they predict the output of the system (in terms of any number of criteria) given the 'input' to its interacting subsystems. For these reasons this text is exclusively concerned with computer-based simulation models which from this point on will simply be referred to as simulation models.

The use of simulation in the study of systems has become standard procedure in many disciplinary areas within agriculture, ranging from natural-resource system planning, through grazing-livestock systems, ruminant metabolic systems, plant physiological systems, to biochemical and cellular systems. This widespread application reflects a number of specific advantages associated with the approach:

1. It enables the study of systems where real-life experimentation would be either impossible, inordinately costly or disruptive.
2. By synthesising systems in model-form, it permits the exploration of systems that do not exist.
3. It permits the study of long-term effects since the time horizon over which a model is run is within the control of the model-builder.
4. It forces those concerned with building the simulation model to examine the system objectively and consequently undertake a thorough and critical review of knowledge concerning the system. The enlightenment that this process provides is often surprising.

A simulation approach, however, is not without its difficulties and dangers. The more obvious problem areas include the following:

1. It tends to be a relatively expensive and time-consuming process. Of course, today there are few types of agricultural systems that have not been subject to analysis and/or experimentation. Whether this has provided sufficient data for successful model-building depends on the system concerned. Nevertheless the process of analysis, assembly of data, model-construction and proving absorbs a lot of costly resources in the form of skilled man hours and computer time. The costs involved and the time taken are frequently underestimated, though as Dillon (1976) points out, the time taken can often be a function of the time available, the implication being that model-builders are prone to continue development of their creation beyond the point where satisfactory implementation could take place.
2. Frequently, data about, and knowledge of, the essential structure of a system will be inadequate; simplifying assumptions about the system structure will nearly always have to be made and compromises in using data from varying sources will be involved together with interpretation of data which have been developed or presented in an inappropriate form.
3. Problems will be encountered in proving (validating) a model (a full account is presented in Chapter 5). It will never be possible to prove a simulation model 'true', yet the use of the computer can lead the unwary to accord the results a greater degree of precision than is justified. The model may contain undetected flaws, poor data transformations and intentional or unintentional biases included by the model-builder. All of these can seriously affect the validity of information provided by the model.

Simulation modelling is not simply a matter of transcribing known formulae into a form suitable for computer analysis. It is rather an art which requires ingenuity, foresight, resourcefulness and, above all, integrity on the part of the model-builder. And, as with other arts, it is how the technique is used, and not the technique itself, which determines success or failure. The purpose of this text is to assist potential model-builders in, and refresh old hands about, the wise application of simulation modelling in agricultural-systems research.

A General Statement of Procedure in Systems Simulation

Systems simulation has two basic phases; one involving the construction of a model and the second concerned with the use of the model in developing or controlling the real system. The whole process of simulation, however, can be usefully considered as embracing a number of interlinked though clearly defined steps. Figure 1.3 illustrates these steps and shows some of the more obvious linkages.

Step 1: *Definition of the system and statement of objectives*
A clear statement of the reasons why the system-simulation work is being carried out is an essential first step. It should be made clear exactly what problems the model will be used to solve and in what degree of detail information is required from the model.

The system is largely defined in terms of the placement of the boundary. A first attempt at the location of the boundary is therefore necessary even if, at a later stage, data availability and/or sensitivity analysis (Step 5) force modifications.

Step 2: *Analysis of data*
Whether or not those involved in the systems research have an overall knowledge of the system under investigation, close attention must now be paid to its essential structure. The systems researcher must examine the system in relation to the objectives laid down in Step 1. His immediate task is to translate his understanding of the system into a basic model-structure—to create a framework on to which he can pin quantitative data. The way in which this is accomplished will influence both the data requirements and the performance of the model. This initial learning stage is a vital one for it establishes the fundamental model-structure.

The model-design is to a large extent dependent on the data available or on the feasibility of generating data within the time limits set by the research. An ideal structure may have to be forsaken because of data limitations. The structure and the quantification of the model are thus intimately bound together; both eventually influencing the effectiveness of the final model.

Step 3: *Model construction*
It is useful at this stage of the work to stop and reflect on the need to proceed

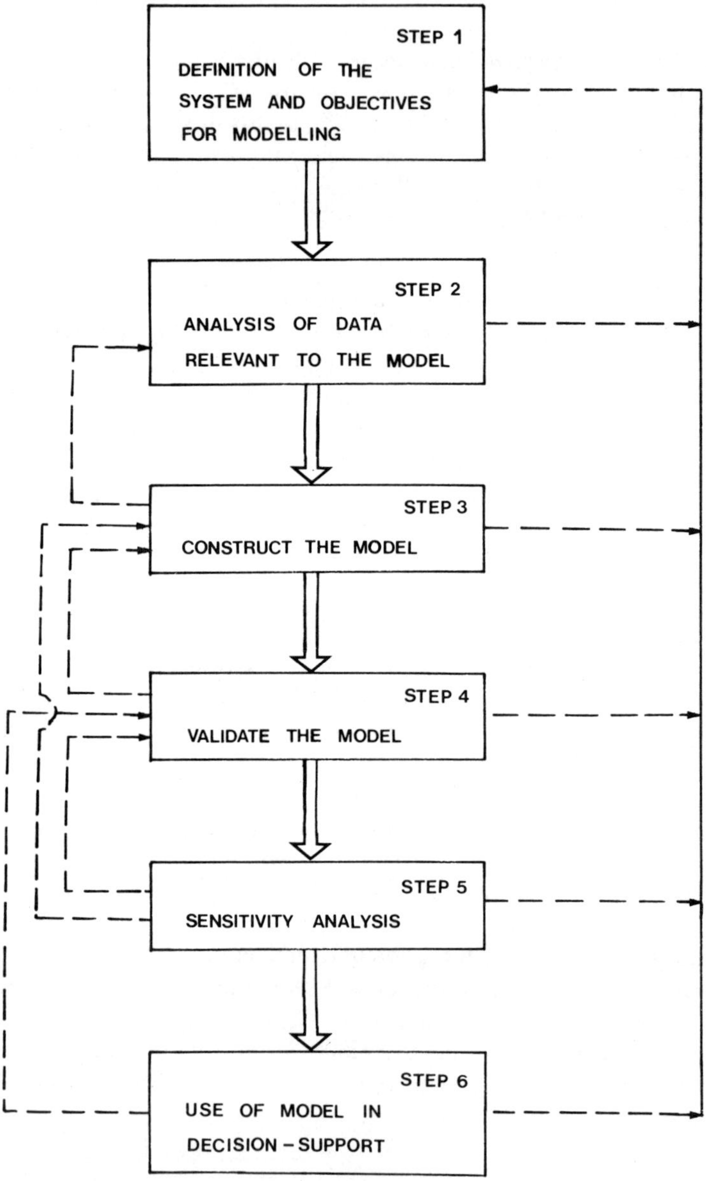

Fig. 1.3. The basic steps of systems simulation.

with the whole simulation process from Step 3 to Step 6. So far we have not committed ourselves to a vast input of time but we have clarified objectives for the study and taken a careful and thorough look at the possibility of meeting these objectives with the information and knowledge to hand. Therefore, we should now ask, is it essential to approach this problem by building a simulation model or are other (simpler) approaches possible? Remember, in any problem-solving exercise, the technique applied should fit the problem rather than the problem become modified to suit the technique. This is the point at which we might best make the decision; before we are totally committed to an extensive modelling exercise.

Assuming the decision to proceed with the simulation model has been made, we move on to the operation of constructing the model itself. This is a multistage procedure which will be detailed in Chapter 2. For now, simply note that the model-builder will not normally attempt an immediate representation of the system under study in the form of a computer program. Rather, he will approach this final model by way of a series of simple symbolic representations in diagrammatic form. First, he may prepare a simple diagram which illustrates the components of the system to be isolated and establishes their relationships with one another. This diagram will be further developed to express the dynamics of the system. Finally, a detailed diagram is constructed which is readily translated to a computer program. These diagrammatic representations are important steps in model-development and construction. Gradually, there will be interplay between the diagrammatic model-form and the computer model-form as data are appended to the diagram and as more dynamic components are included. This interplay gradually establishes the balance between model-structure and model-detail: usually some sacrifice of what might be envisaged as ideal structure will be forced by data reality. This procedure emerges as data needs are established and quantification of relationships previously expressed in diagrammatic form takes place, hence the feedback indicated in Fig. 1.3 between Step 3 and Step 2. The final symbolic model will be expressed as a series of relationships and equations in a computer language.

Selection of the most suitable computer language is a major decision. While the range of languages is large, in practice the selection is greatly restricted by the skills of the model-builder or by the availability of alternative languages on the immediate computer facility. A key decision will be whether to use a special-purpose simulation language or a general-purpose language. Special-purpose languages have particular devices which can assist the computer implementation of simulation models. For example, they often have

1. devices which assist in representing the dynamics of systems;
2. easily used input–output procedures with emphasis on provision for graphic output;
3. facilities to assist in error detection.

In addition, each will have devices that are helpful in representing some particular type of system. Against these features, general-purpose languages are more flexible and can be used to represent a wide range of different types of system. This discussion will be expanded in Chapter 3 but for now it should be noted that the authors believe, on balance, a general-purpose language is most suitable for the apprentice model-builder (at least) except where particular circumstances force the use of a special-purpose simulation language. To this end, examples in this book will all be presented in FORTRAN which is a widely understood general-purpose language.

Step 4: Validation of the model

Validation is a particularly awkward step in the simulation process. Obviously it involves a testing and an assessment of the model created in Step 3. Recall that the model has been built for some precisely specified reason(s) and it is in relation to this reason(s) that we should both build and judge the model. The model needs to mimic the real system sufficiently well to fulfil the purposes for which it was developed. The use to which the model will be put and the purposes for which it was designed will be the framework within which we must make the assessment of model validity.

When we are confident that the behaviour of the model is satisfactory, the formal validation process is over. Although a formal process of validation is always recommended, the process of gaining confidence in the model is generally a slowly emerging one over the period of model-construction, through formal validation, to application of the model. During this time, assessment and modification will proceed, essentially two sets of judgements being made on the way:

1. that the model is not different from the real existing system to a degree that will detract from the value of the model for the purposes for which it was designed;
2. that if the model is accepted as being adequate then the decisions made with its assistance will not be measurably less correct than those made without the benefit of the model. Such an *ex ante*

assessment is extremely difficult and cannot be other than subjective in nature. Subjective judgements in this regard are, however, not without value and this aspect of validation may often be the most relevant.

It is because of the difficulties associated with these two sets of judgements that validation remains an elusive issue in the simulation procedure. Numerous exact statistical tests are available to determine whether the model-behaviour is different from real-system behaviour to some stated level of significance. The precision of such tests should not overshadow the conceptual problems in applying them. There is no mechanical procedure that permits the conditions for acceptable validation by a single (or even a series of) statistical comparison(s) of the model-performance against some recorded or measured data from an existing system which the model represents. Any statistical tests that may be carried out and which are positive in favour of the model add to the model-builder's confidence that his creation is functioning well. The acceptable level of confidence is achieved normally by a series of assessments and subsequent modifications until such time as the model is to be applied in support of decision making. In this regard the feedback indicated in Fig. 1.3 between Step 4 and Step 3 is important.

Step 5: Sensitivity analysis

Sensitivity analysis is a procedure carried out on the completed and, at least partly, validated model which involves exploring the operation and performance of the model. In successive 'runs' of the model under identical environmental conditions, the value of a parameter may be changed. The resultant modification in model-output will be analysed to determine whether or not the changed parameter values are of material consequence. A sensitive parameter is one which causes a major change in model-output; the model is said to be sensitive to such a parameter. A similar, though in practice more complex, procedure can be envisaged to isolate sensitive subsystems and relationships. Sensitivity analysis should, of course, be carried out with respect to the model's purpose. If sensitivity to a given parameter is not important as far as the use of a given model is concerned, then there is little need to try to account for the sensitivity.

The mechanism for sensitivity analysis will be explored later in the book but we should note some important uses at this point.

1. If the model is a good representation of the real system, a sensitive

region established in the model can, by association, be considered to be so in the real system. This leads into implementation of model-results since, in the management of the real system, such sensitive regions are clearly points where close control is needed if the operation of the system is to be directed effectively. Isolation of control points by sensitivity analysis on the model provides direct guidelines for management.

2. If a model is shown to be sensitive to particular aspects of its design or of its data about which suspicions of inaccuracy or inadequate knowledge linger in the model-builder's mind, then those parts of the model must now be closely examined. Close validation may be warranted on specific subsystems, relationships or individual parameters. This may well reveal an overemphasis on this particular part of the model. If this is so, modification of the model should take place aimed at establishing a more balanced model of the real system. In this respect sensitivity analysis is providing guidelines for validation as indicated in Fig. 1.3 in the feedbacks between Step 5 and Steps 3 and 4.
3. An established sensitive area or element in a system as isolated by sensitivity analysis on the model clearly should be well understood and properly represented in the model. To the extent that 'good' knowledge of these sensitive points should be available and be built into the model, any further systems analysis in the form of scientific experimentation should be directed towards them. Several such points may emerge in sensitivity analysis and a ranking of them in relation to the degree of sensitivity of model-output is a first step in providing priorities for a conventional research programme associated with the system being modelled.

Reference to Fig. 1.3 indicates the potential importance of sensitivity analysis in the whole process of systems simulation. This phase represents a major learning period for the model-builder with implications for formal validation, model-structure and detail and indeed in the possible redefinition of the objectives of the study. It will hopefully be a period of growing confidence in the model even if it involves modification to the model. Experience indicates that sensitivity analysis is an on-going procedure even when the model is finally approved for application. Application may indeed be considered as an extension of sensitivity analysis, involving as it does the 'running' of the approved model under alternative rules.

Step 6: Application of the model
The systems-simulation procedure is aimed at decision-support in relation to the real system under consideration. The exact nature of this will depend very much on the type of system involved. Broadly two main areas of application can be delineated which are closely interlinked. Basically, the two areas are in directing further systems analysis and in assisting management and development of the real system. For some systems the later application will not have immediate relevance: those involved in fundamental biological research will often be concerned with such systems. Here, the major objective of modelling will be to assist in the closer understanding of the system and to use sensitivity analysis to guide the research programme.

For more applied systems which demand management control, the provision of information from the model which will provide a basis for decision making can be the most important use. When we are involved with the management of farm systems there are two possible approaches to decision-support:

1. In a normative sense, a model can be constructed representing a typical farm system. Such a model can be used by advisory personnel in order to demonstrate to farmers the way this typical system is likely to respond to various alternative management policies. Provided individual farmers can associate their own farm system with the typical system represented in the model, much is to be gained in a general sense.
2. In a direct decision-support role, a tailor-made model of an individual farming system can be of great value in the assessment of management options and in the control of the system.

Obviously the model of the representative farm is of limited value in providing clear planning and control information to the individual farmer for his farm system. On the other hand, to provide a detailed simulation model of an individual farm system would be inordinately expensive. Reconciliation of these two extremes of application is fully discussed in Chapter 7.

Concluding Remarks

This chapter has developed the concept of the computer-based simulation model. The objective has been to present a brief but comprehensive review of the purpose and place of systems research as a scientific discipline. The ideas introduced in the preceding pages will be developed in full in the chapters to come. However, it is important that the reader has at least an elementary comprehension of the principles of systems research before launching into the detailed concepts. It may, therefore, be useful to pause at this stage and briefly review the ideas which have been presented before progressing further. The reader may also find it useful to re-read this chapter periodically in order to refresh his or her memory on the overall concepts of systems research.

Appendix

Two models provide many of the examples used in the text to illustrate particular points of simulation methodology: they will further be employed in appendices to certain chapters to reinforce concepts and methodology set out within the chapters. The first model represents growth and development of the disease *Puccinia hordei*, or leaf rust of barley, on the growing barley crop and is intended for application in assessing different disease-management strategies in a growing crop of barley. The second model is a generalised model of a pig-production unit and can provide the detailed physical and financial outcome of any selected management policy on a given pig herd. While the first model has a far more detailed biological base than the second, both models have features which make them useful for the teaching, as well as the practice, of simulation. A brief outline of the systems to which these two models relate is provided in this appendix as well as some introductory statements about the models. Further detail will be introduced in later appendices.

THE BARLEY-LEAF-RUST MODEL

Plant diseases were originally studied because of the losses they caused in crops. However, a paradoxical situation exists today where there are few reliable estimates of crop loss on a regional scale, or the amount of loss caused by a disease on a specific crop (James, 1974). This lack of disease-loss information does not undermine the reality that plant diseases are still a major constraint on world food production. A knowledge of the amount of loss due to plant diseases can be used to help orientate research priorities as well as for deciding on the need for disease control. A more active use of this was proposed by James (1974) through incorporating a quantitative disease-loss relationship in a disease-control scheme. Such a scheme would seek to rationalise the use of each fungicide spray on the basis of the potential (extra) amount of revenue (yield) it would generate. The need to rationalise fungicide use for crop protection on an individual farm is related to profit-maximising objectives and is based on the economic concept of marginal value product for the farm input. When the single farm situation is extrapolated to a national and international scale, rational fungicide use assumes the roles of minimising energy demand and chemical pollution.

The type of decision a farmer can make to maximise profit in a cropping system constrained by disease is strongly influenced by several factors. Risk aversion may make him avoid a specific crop because experience in his locality shows it to be repetitively disastrous, or he may select a particular variety because of its proven resistance to a key disease. A farmer may make use of a certain rotation system to reduce inoculum potential of a disease or he may modify his drilling operations in anticipation of disease infection at some point in the cropping season. However, once he has affirmed his decisions to plant a given crop, the farmer has created the agroecosystem (Harper, 1974) within which he hopes to reap maximum economic benefit through his managerial expertise. If he believes, or is afraid, that in spite of his actions so far his crop will still be subject to disease, he may opt for a calendar spray schedule as an insurance without consideration of the cost-benefit relationship of each spray he applies and without feedback knowledge of the status of disease. The later in the season a farmer is forced to make his decision, the more limited will be the range of options available to him. In situations where fungicide cost is high and where the disease is not regularly damaging, the farmer would be expected to monitor the presence of disease and conceivably the level of the disease attack. He then makes, whether consciously or not, a management decision to spray based on his expectations of disease and of yield reduction. There is as yet little technical ground on which this last action—the disease-management option—can be taken by a farmer except for disease-forecasting schemes based on lengths of favourable weather for disease development (Krause & Massie, 1975). Such schemes generally forecast the likelihood of disease infection and assume that crop loss would be high in the absence of control. A system model which simulates the natural disease epidemic under a given set of environmental conditions, estimates the amount of yield reduction due to this epidemic, and estimates the amount of yield increase with each control measure, will eliminate much of the subjectivity involved in the disease-management option.

THE PIG-HERD MODEL

Intensive livestock enterprises such as modern pig production rely for their profitability on high levels of production. Significant and costly inputs of capital, feed and energy are required to achieve these production targets and small variations in the costs or the quantities used of such inputs can have marked effects on enterprise profitability. To assist the farmer in monitoring the flow of cost and production from his pig unit, an information system which includes a simulation model of a pig enterprise has been developed. This model is sufficiently general to accommodate all feasible systems of production, allowing in each case for the correct phasing and timing of inputs and outputs. The boundary for this model is defined as the pig enterprise and the cash flows generated by this enterprise. No attempt is made to incorporate the model within the whole-farm boundary by inclusion of such effects as tax allowances, farm debt servicing, or interest on surplus cash. This is not to say that the model needs to be entirely divorced from the remainder of the farm system. The model is tailored to the individual unit by the data input so there is considerable flexibility in the manner in which a manager may specify the complete model for his unit.

The model can be used initially to derive a suitable annual plan for the pig unit. Key parameters (targets) from this plan are then saved on a computer file. Data are regularly collected from the farm and sent into the information system for processing. The actual results from the plan are compared against the targets necessary to achieve it and a report returned to the farmer indicating how the enterprise is performing relative to the predetermined plan.

The pig-herd model is part of an information 'package' which provides the individual farmer with the facilities for budgetary control normally outside the scope of the small business.

References

Ackoff, R. L. (1973). Science in the systems age: beyond IE, OR and MS. *Operations Research*, **21,** 661–71.

Anderson, J. R. (1974). Simulation: methodology and application in agricultural economics. *Rev. Mktg Agr. Econ.*, **42,** 3–55.

Churchman, C. W. (1971). *The Design of Inquiring Systems: Basic Concepts of Systems and Organisation.* New York: Basic Books.

de Wit, C. T. (1970). Dynamic concepts in biology. *In:* Setlick, I. (ed), *Prediction and Measurement of Photosynthetic Activity.* Wageningen: Centre for Agricultural Publishing & Documentation.

Dillon, J. L. (1976). The economics of systems research. *Agr. Systems*, **1,** 5–22.

Harper, J. L. (1974). Agricultural ecosystems. *Agroecosystems*, **1,** 1–6.

James, W. C. (1974). Assessment of plant diseases and losses. *Ann. Rev. Phytopathol.*, **12,** 27–48.

Krause, R. A. & Massie, L. B. (1975). Predictive systems: modern approaches to disease control. *Ann. Rev. Phytopathol.*, **13,** 31–47.

Mihram, G. A. (1972). *Simulation: Statistical Foundations and Methodology.* New York: Academic Press.

Shannon, R. E. (1975). *Systems Simulation: The Art and the Science.* Englewood Cliffs: Prentice-Hall Inc.

Teng, P. S., Blackie, M. J. & Close, R. C. (1978). Simulation modelling of plant diseases to rationalize fungicide use. *Outlook Agr.* (in press).

Throsby, C. D. (1973). *New Methodologies in Agricultural Production Economics: A Review.* Macquarie University, School of Economic & Financial Studies, Research Paper No. 29, Mimeo.

Wright, A. (1971). Farming systems, models and simulation. *In:* Dent, J. B. & Anderson, J. R. (eds), *Systems Analysis in Agricultural Management.* Sydney: Wiley.

Further Reading

Anderson, J. R. (1972). Economic models and agricultural production systems. *In: Proc. Australian Soc. Animal Prod.*, **9,** 77–82.

Beiston, R. J. & Peters, G. (eds) (1972). *Systems Behaviour*. London: Harper & Row.

Buckley, W. (ed) (1968). *Modern Systems Research for the Behavioral Scientist*. Chicago: Aldine.

Charlton, P. J. & Thompson, S. C. (1970). Simulation of agricultural systems. *J. Agr. Econ.*, **21,** 373–84.

Churchman, C. W. (1971). *The Design of Inquiring Systems: Basic Concepts of Systems and Organisation*. New York: Basic Books.

Cleland, D. I. & King, W. R. (1975). *Systems Analysis and Project Management*. New York: McGraw-Hill.

Conway, G. R. (1973). Experience in insect pest modelling: a review of models, uses and future directions. *Mem. Ecol. Soc. Australia*, **1,** 103–30.

Crom, R. J. (1975). Development of a systems approach for livestock research in ERS. *Am. J. Agr. Econ.*, **57,** 509–12.

Daetz, D. & Portell, R. H. (eds) (1974). *Environmental Modelling: Analysis and Management*. New York: Dowder, Hutchinson & Ross.

Dalton, G. (ed) (1975). *Study of Agricultural Systems*. London: Applied Science.

Dent, J. B. & Anderson, J. R. (eds.) (1971). *Systems Analysis in Agricultural Management*. Sydney: Wiley.

Forrester, J. W. (1969). *Principles of Systems*. Cambridge, Mass.: Wright-Allen.

Geier, P. W. & Clark, L. R. (1976). On the developing use of modelling in insect ecology and pest management. *Australian J. Ecol.*, **1,** 119–27.

Greenblatt, C. S. (1972). Gaming and simulation in the social sciences: a guide to the literature. *Simulation and Games*, **3,** 477–91.

Heady, E. O. & Candler, W. (1958). *Linear Programming Methods*. Ames: Iowa State University Press.

Jeffers, J. N. R. (ed) (1971). *Mathematical Models in Ecology*. Oxford: Blackwell.

Jones, J. G. W. (ed) (1970). *The Use of Models in Agricultural and Biological Research*. Hurley: Grasslands Research Institute.

Jones, J. G. W. (1974). Models of agricultural production systems and their application. *In: Economic Factors Affecting Egg Production*. Proceedings of the Tenth Poultry Science Symposium, 18–20th September 1974, Edinburgh: British Poultry Science.

Mihram, G. A. (1972). *Simulation: Statistical Foundations and Methodology*. New York: Academic Press.

Nott, S. B. (1977). Dairy farm planning and systems analysis. *J. Dairy Sci.*, **40,** 464–72.

Optner, S. L. (1968). *Systems Analysis for Business Management*. Englewood Cliffs: Prentice-Hall Inc.

Optner, S. L. (ed) (1973). *Systems Analysis*. Harmondsworth: Penguin.

Orcutt, G. H. (1960). Simulation of economic systems. *Am. Econ. Rev.*, **50,** 897–907.

Patten, B. C. (ed) (1975). *Systems Analysis and Simulation in Ecology*. New York: Academic Press.

Rose, C. W. (1973). The role of modelling and field experiments in understanding complex systems. *In:* Bofinger, U. S. & Wheeler, J. L. (eds), *Developments in Field Experiment Design and Analysis*. Commonwealth Bureau of Pastures & Field Crops, UK, Bulletin No. 50.

Russell, C. S. (1975). *Ecological Modelling in a Resource Management Framework*. Washington: Resources for the Future.

Russell, J. S. (1975). Systems analysis of soil ecosystems. *In:* Paul, E. A. & McClaren, A. D. (eds), *Soil Biochemistry*, Vol. 3. New York: Marcel Dekker.

Shannon, R. E. (1975). *Systems Simulation: The Art and the Science*. Englewood Cliffs: Prentice-Hall Inc.

Siaty, T. L. (1959). *Mathematical Models of Operations Research*. New York: McGraw-Hill.

Tummala, R. L., Haynes, D. L. & Croft, D. (eds) (1976). *Modelling for Pest Management: Concepts, Techniques and Applications*. Michigan State University.

Watt, K. E. F. (ed) (1966). *Systems Analysis in Ecology*. New York: Academic Press.

2

Model-Construction

Getting Started

One of the most basic—yet rarely discussed—problems in any scientific research is where to begin. The systems researcher will invariably be beset by this difficulty and a model which is poorly conceived will inevitably fail to produce satisfactory results. A good beginning is a prerequisite for a worthwhile model.

The first step in getting started is to formulate the problem and to assemble the required information. In many simulation studies, the model-builder will not be the ultimate user of the model. Therefore the model-builder must determine right at the outset that he fully understands the problem in the same context as the user. The user will probably be fully conversant with the background of the particular problem and may well make conscious or unconscious assumptions as to the model-builder's knowledge. Often there may be several users and also the model-building may be a group exercise. Establishing a common understanding between all parties requires tact, understanding and experience. The model-user will describe the problem as he understands it, complete with his own biases and assumptions. It is the model-builder or analyst's job to distinguish between facts and opinions—both are important for his work but must be handled differently. In like view, he must be able to identify his own prejudices and must beware of jumping to conclusions based solely on past experience.

Where there is a real system to study, the analyst should observe the real system in the field as much as practicable so that he becomes fully conversant with the behaviour of the existing system and the jargon used to describe it. This learning and observing phase is of considerable importance in 'gaining a feel' of the system and of the problem to be solved.

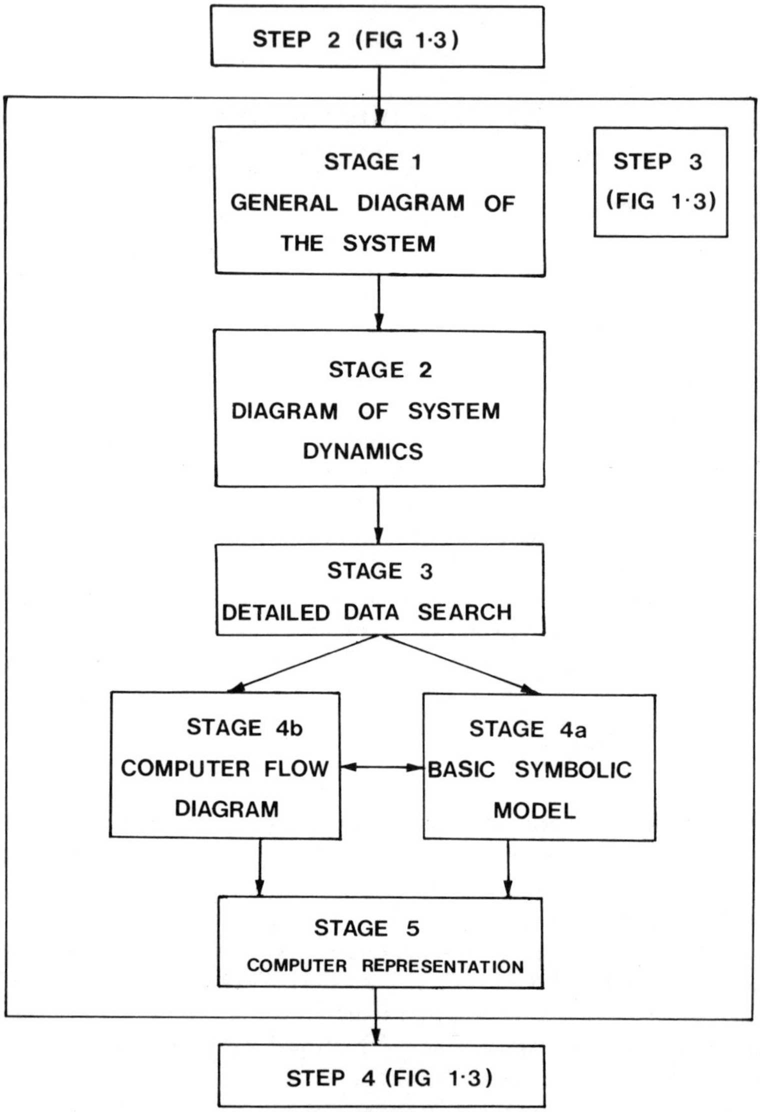

Fig. 2.1. *Basic stages in model-building.*

The Process of Modelling

At the start we should state that there is no set procedure involved in constructing simulation models. The objective is to work towards a functioning model as efficiently as possible. How this is accomplished will depend on the experience of the model-builder, the complexity of the system and his familiarity with the system. Figure 2.1 provides a framework with which to consider the various stages: it is not claimed to be exhaustive but rather is illustrative of the procedures that many modellers find convenient. Refer back to Fig. 1.3 and it will be clear that Fig. 2.1 is an expansion of Step 3 in that figure. Step 3 now is broken down into a series of stages, each of which is described below.

Stage 1

Stage 1 is the first creative stage in which the general idea is to promote systematic and clear thinking about the system under study and to set the component parts in appropriate juxtaposition. A pictorial representation of the system is useful and there are a number of ways this can be done. One possibility is a block diagram which is a set of boxes indicating the major components and linking arrows to indicate an interrelation, or a flow of material or information. Information flows are often presented as dotted lines but again no formal distinction is necessary.

Figure 2.2 is a block diagram setting out an initial stage of modelling for a system concerned with the growth of a grazing animal. This first presentation is general and is concerned with identifying only the major subsystems; hence subsystems referring to the feeding value of the herbage produced, to the intake of forage by the ruminant or to the metabolic processes of the animal are not specified. These subsystems are, however, implied in the interconnecting arrows and would appear in later stages. The block diagram illustrated in Fig. 2.2 simply shows that liveweight gain in ruminants is related to the starting weight of the animal, its feed intake and the climate. Feed intake in turn can be divided into feed from grazing and from supplementary feed, and so on.

Stage 2

Early in the development of the model it is essential to begin consideration of the essential dynamics of the system. The most important characteristics of a system stem from its dynamics, and the rates of change of components of the system form a major focus of attention. It follows that any model of a system must properly represent both rates of change and also factors which influence, or which are influenced by, these rates.

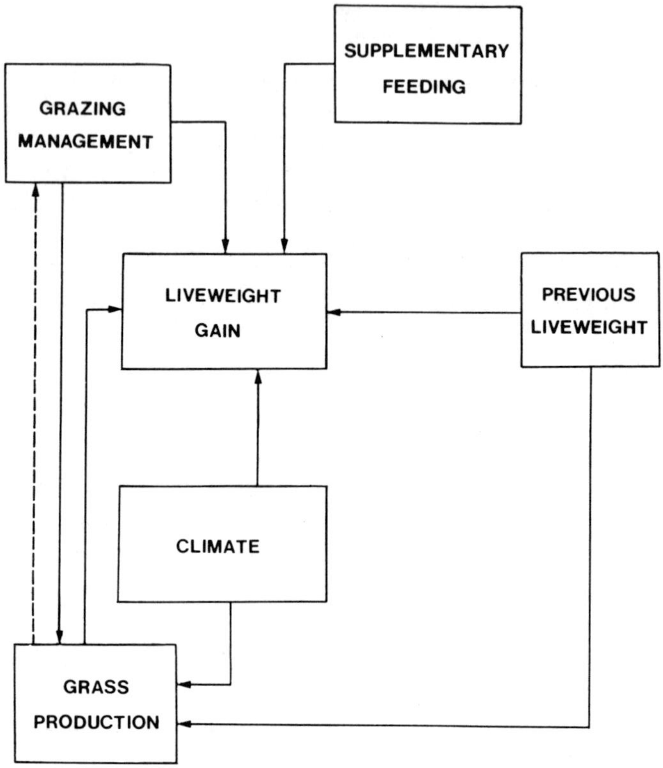

Fig. 2.2. Block diagram for a grazing system.

Consider the soil-water–plant-growth system; our interest will be centred *inter alia* on the following rates of change:

1. the precipitation rates;
2. the irrigation rate (if any);
3. the actual evapotranspiration rate;
4. the potential evapotranspiration rate;
5. the rate of plant growth;
6. the rate of uptake of nutrients; and so on.

In the second stage of model-building we begin to define the essential rates of change and the dynamics of the system. At this stage of our work we will still be concerned with a diagrammatic model but more sophisticated symbols may be used. Perhaps the most helpful notation is that established by Forrester (1968) from his studies of industrial systems and presented in his book '*Industrial Dynamics*'. We need only consider four such symbols from a fuller set described by Forrester.

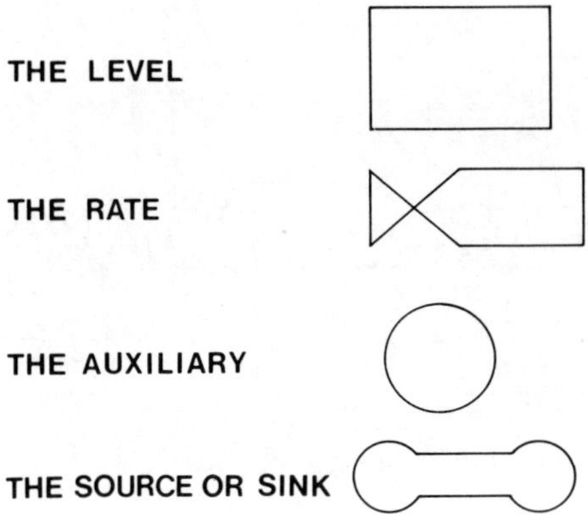

The use of such symbols is quite simple: the *level* is a major component of a system with whose changing value we are particularly concerned; the *rate* directly brings about change in the value of the level, while the *auxiliary* represents a factor which may influence rates (i.e. may change the value of a rate) or may be influenced by a level or a rate. An easy way to envisage the application of these symbols is the example of a bath tub being filled with water while a group of children are splashing about in the tub with the clear intention of emptying it before the soap can be applied. Figure 2.3 sets the unruly scene. The amount of water in the tub is changing continuously; increasing by the inflow from the tap and decreasing by the outflow caused by the splashing children. Figure 2.3 is a rather cumbersome (although visually attractive!) method of representing the bath-tub system (it should, though, be recognised as an iconic model) and can be considerably simplified as shown in Fig. 2.4.

At a more serious level, a section of a soil-water–plant-growth system is presented in *Industrial Dynamics* format in Fig. 2.5. In this diagrammatic model of the subsystem concerned with the soil-moisture budget, the level of soil moisture is depleted by actual evapotranspiration and increased by rainfall. The level in the model is the available soil moisture while the two rates are useful rainfall and actual evapotranspiration. Actual evapotranspiration is seen to depend on potential evapotranspiration and the maximum rate of water uptake; in fact when there is sufficient soil moisture to be non-limiting to plant growth, actual evapotranspiration may be set equal to potential evapotranspiration and when soil moisture is limiting actual evapotranspiration is set equal to the maximum rate of water uptake by plants which in turn is dependent on the available soil-moisture level.

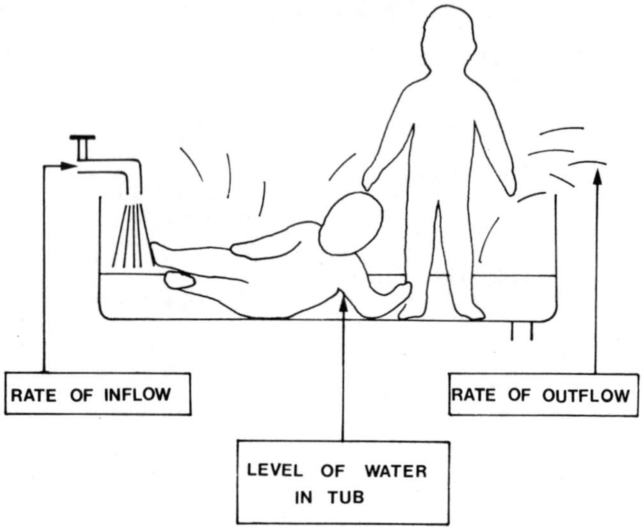

Fig. 2.3. First concepts in the dynamics of modelling.

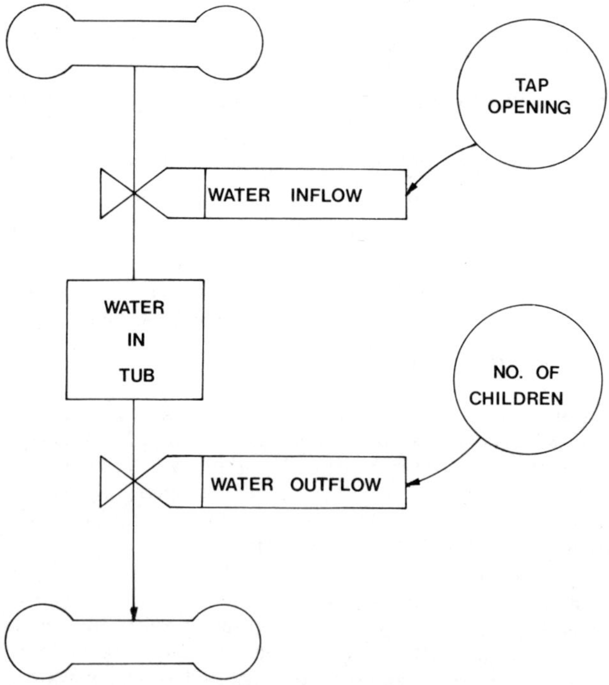

Fig. 2.4. Use of Industrial Dynamics *symbols.*

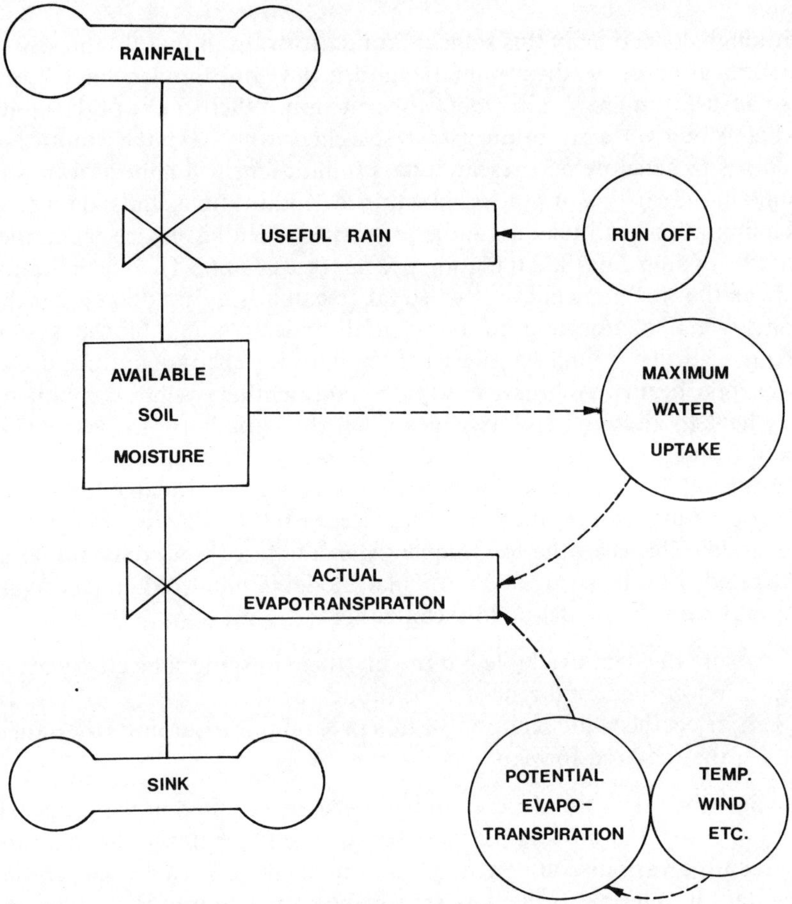

Fig. 2.5. Industrial Dynamics *symbols depicting dynamics of available soil moisture.*

The use of the simple *Industrial Dynamics* symbols forces a representation of the system's dynamics and leads directly to the first stage of the symbolic model as we shall see in Stage 4b. It may well be that, for the experienced modeller, the *Industrial Dynamics* format can replace the simple block diagram in Stage 1 but whether as a first-stage or second-stage diagram, the *Industrial Dynamics* presentation will be found a most valuable step in the modelling process.

Stage 3

Although conceptually this stage is straightforward in that it represents a quantification of the diagrammatic models developed under Stage 2, it is also a frustrating period in model-construction. Whether or not the model is being built in association with researchers who have the ability and facilities to generate new research data, limitations will almost always be found in the existing or potential (within acceptable time limits) data base. The limitations will take a number of forms which have been mentioned broadly in Step 2 of the simulation process (see Chapter 1). One limitation may be that the model developed so far (recall it is only a diagram at this stage) while conforming to our initial understanding of the system frequently will be found wanting when the hard search for quantified theory (concepts) begins. Alternative ways of representing system components may have to be found and may mean that the initially prescribed level of detail for certain parts of the model will have to be changed.

A second limitation may be that although data are available for a given area of the model, they may be unsuitable for the specific requirements of the model. This is a situation frequently encountered since data are rarely generated with the needs of a particular model in mind. When faced with this situation the model-builder can either

1. use the data available but modify them in some subjective way, or
2. reject them and redesign the model, or
3. reject them and set up a research programme to generate the data in the required format.

The decision taken will depend on the precision required in the particular component of the model (and the obvious need to balance the precision between the various components), the time factor involved in generating new data in relation to the date set for the completion of the model and finally on the resources available either to make effective subjective modifications to inappropriate data or to generate new data. A final limitation with respect to data that the modeller will encounter is that the suitable data for specific relationships or rates of change are simply not available.

For convenience in experimentation, lactation data on dairy cattle tend to be collected from lactation peak onwards. Yet, for modelling, a vital piece of information is milk yield from parturition to peak yield. In spite of massive research into dairy production, this small but essential piece of information is absent from the literature and hinders modelling of the milk-production process.

The option of subjectively estimating data is a new concept in our discussion and since it is used with reasonable frequency it should be examined carefully. (We should note that this option is also available to the modeller who has data in an inappropriate format.) The decision to estimate values subjectively is not such a wild procedure at this stage in our model-development as might be thought and it could save entering into an experimental programme with all the costs in time, facilities and resources this implies. Assume that the difficulty lies with a single piece of information—a coefficient representing a rate-of-change constant. We will also assume that we have only the broadest notion of the value of this coefficient. The procedure would be as follows:

1. Determine the best 'guestimate' for the coefficient.
2. Proceed with the development of the rest of the model and complete its structure and its implementation on the computer.
3. Establish appropriate parameters for monitoring the performance of the whole model.
4. 'Run' the model with the decided value for the unknown coefficient—note the values of the performance parameters.
5. 'Run' the model several times with different, widely-spaced values of the unknown parameter each time and without changing any other aspect of the model noting the values of the performance parameters.
6. Compare the values of the performance parameters for each 'run'.
7. In the case where different assumed values of the unknown coefficient do not markedly influence the performance parameters we may say that the model is not sensitive to the selected value for this particular coefficient. Our judgement might then be that the 'guestimate' is acceptable and may, without loss of value to the model as a whole, be incorporated in its structure.
8. In the case where markedly different output parameters are consequent on the different assumed values of the unknown coefficient we would say that the model is sensitive to this particular coefficient. The 'guestimate' should now be used with caution and it would be advisable to extend the data search perhaps by way of a new research programme. Alternatively, it may be possible to restructure the model in such a way as to reduce the sensitivity of the coefficient.

The pig model provides a useful example of 'guestimating' a key model-parameter. Breeding sows are expensive to maintain and, therefore, managers of breeding-sow herds try to ensure that as many of the sows in the herd as possible are either pregnant or lactating. The non-productive stage of a sow's life is that period from when her piglets are weaned until she is successfully mated. The first oestrus or 'heat' period occurs about 1 week after weaning and thereafter at 3-weekly intervals. Sows can only be mated during oestrus periods. In mimicking the behaviour of the breeding unit, the pig model had to estimate the proportion of sows which were successfully mated at each oestrus period. Reliable farm data on this parameter were unavailable but farmers were able to specify reasonably accurately the number of litters each sow produced on average in a 12-month period (this is termed the 'farrowing index'). It was this figure which was used in the model to estimate the numbers of sows successfully mated at each oestrus period. Say the farmer weaned pigs at 5 weeks of age and had a farrowing index of 2·0 litters per sow per year. If all his sows conceived on the first oestrus period and the gestation period for sows is 17 weeks, his potential farrowing index would be 2·29. The difference between the actual and the potential farrowing indices is caused by sows only becoming pregnant after a second or third mating.

Therefore, in the pig model, the following assumptions are made:

1. The number of sows successfully mated at each oestrus period is based on the ratio between actual and potential farrowing index.
2. Sows not pregnant after three matings are classified as culls and replaced.

While this approach was fairly crude in a biological sense, it did result in a satisfactorily accurate simulation of breeding-herd behaviour using data that were readily available and reliable.

We will be returning to the theme of application of sensitivity analysis at a number of points in the book and its general potential in guiding further analytical research has already been mentioned (Chapter 1). Clearly it is pointless and wasteful to gear analytical research to generate data which, although they may be filling a gap in our knowledge, do not materially assist in the current modelling work. However, the potential for guiding analytical research in a problem-solving mode by way of a simulation model and sensitivity analysis should be noted and underlined: we will return to this theme again in Chapter 7.

In general terms the process of quantifying the conceptual base of a model and refining this quantification will proceed from this stage through to the final application. It should be seen as a continuous process involving interaction with other steps in construction of the simulation model and receiving final guidance from the completed model.

Stage 4

This stage of model-construction involves preparing the model for computer coding. The model at this point is defined in terms of its overall structure and general data. It is now necessary to get into the detail of the model so that a comprehensive picture of the eventual model can be built up. One major factor in this process is the way in which time is to be built into the model. If time is advanced on a fixed unit basis, the model is of a 'time-stepping' type wherein time moves forward with the same increment

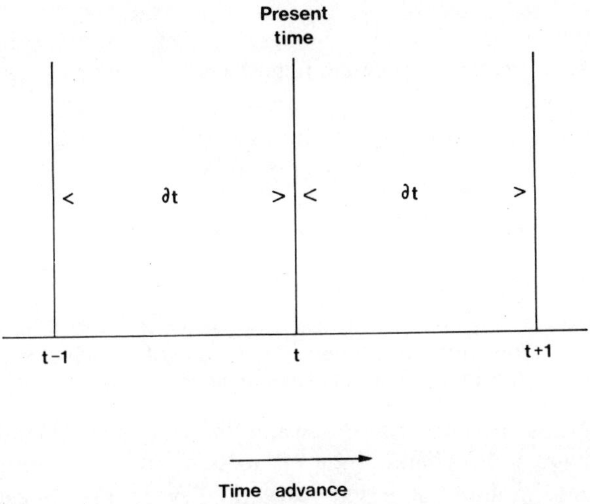

Fig. 2.6. *Time notation in time-stepping simulation.* (*After Forrester*, 1968.)

from the start of the simulation to the end. Events occur and rate changes take place only within this fixed time period. Figure 2.6 sets out the basic time notation used in a 'time-stepping' simulation. The present time is defined as t while at a fixed interval (∂t) on either side of t, $t - 1$ represents the time one interval previous to t and $t + 1$ represents the time one interval later than t. At time t various calculations are carried out in the model and then, when these are complete, time advances one interval (to the right in Fig. 2.6). When this happens, what was $t + 1$ becomes t and what was t becomes $t - 1$ and a further set of calculations are completed before the next step in time is taken. Time is advanced by single units of the predefined size.

Both the pig model and the barley-leaf-rust model are time-stepping models. In the pig model, the time steps are increments of 1 week and the

FORTRAN representation of the annual pig-breeding cycle is summarised below:

```
          DO 100 ITIME = 1, 52
C COMMENCE WEEKLY SIMULATION OF PIG HERD
C VARIABLE 'ITIME' REPRESENTS THE WEEK OF THE YEAR
C IN WHICH EVENTS OCCUR. MINIMUM TIME STEP IS 1 WEEK
                              .
                              .
                              .
                              .
                              .
          (simulation program follows here)
                              .
                              .
                              .
                              .
                              .
C END OF LOOP PERFORMING SIMULATION OF PIG HERD
C
        100 CONTINUE
```

This representation takes the form of a loop in which the variable ITIME is incremented in units of one from an initial value of 1 to final value of 52. ITIME, therefore, simulates a weekly time clock.

The alternative method of representing time may be considered to be an 'event-stepping' model wherein time advances by irregular intervals related to the point in simulated time that succeeding events take place. From the point at which one event occurs, time skips to the point when the next event occurs. This is a procedure commonly used in simulating systems comprising queueing situations. In such situations it may only be necessary to consider time in terms of arrival events and service intervals. Each event is then represented as an instantaneous occurrence in simulated time and is scheduled to occur by the dynamics of the model. An executive routine automatically sequences scheduled events so that they occur properly in simulated time. This scheme may prove more cumbersome to programme since the time-keeping executive routine has to be provided with full details of the time-ordered event sequence. However, there are special-purpose languages (e.g. SIMSCRIPT, GASP) to assist in this type of modelling and these are useful where events do not occur in a regular manner. (See Chapter 3 for a full discussion on special-purpose languages.)

The pig model and the barley-leaf-rust model have event-stepping modules. (The concept of a 'module' will be expanded in Chapter 3; for now it is sufficient to regard a module as a part of the model concerned with some specific task.) In the case of leaf rust, a spore which successfully germinates on a leaf then enters a latent period before forming an infectious pustule. Field experiments conducted alongside the modelling exercise showed that the length of the latent heat period was a function of ambient temperature and the density of infective pustules already on the leaf. Ambient temperature and pustule density were both related in some fashion to the time of year. The event of pustule eruption takes place not on a fixed day of the year but between 4 and 15 days after successful spore germination. Therefore once the leaf-rust model has simulated the successful germination of any spores on a given day of the crop growing season, those spores then enter an event-stepping module. The simplified example below in FORTRAN discusses the operation of this module.

The array LATENT, which has 15 elements, contains groups of latent pustules of different ages. LATENT (1) has that group of spores which germinated in this day of the simulation while LATENT (8) has that group which germinated 8 days ago. The variable IRUPT is the endogenously-calculated length of the latent period or germination delay (some value between 4 and 15 days). Spores germinate and are placed in LATENT (1) by the model. In the next time step of the model (1 day in this case), this group is moved to LATENT (2) and the new day's group of spores placed in LATENT (1). Similarly all other groups of spores are moved up one position in LATENT—the model thus simulating the ageing of latent pustules by 1 day. The group of spores in LATENT (IRUPT) become infectious and are transferred into position 1 of the array INFECT. The action of this module is controlled by the irregular event step IRUPT and is only indirectly influenced by the main week time clock of the model. Figure 2.7 illustrates the operation of this event-stepping module.

It is neither possible nor desirable to present fixed rules for the selection of time-keeping method. The choice will be dependent on many factors—programming ability, computing facilities (both hardware and software), personal preference, etc.—but the following guidelines presented by Shannon (1975) may be of value. The fixed-time-increment method should be considered when

1. events occur in a regular and fairly equally-spaced manner;
2. a large number of events occur during one simulated time interval and the mean length of events is short;
3. the exact nature of significant events is not well known, such as in the early part of a study.

The event-stepping method is useful when

1. the system is static over long periods of time;
2. there is difficulty in determining the size of time increment;
3. events occur unevenly in time and/or the mean length of events is long.

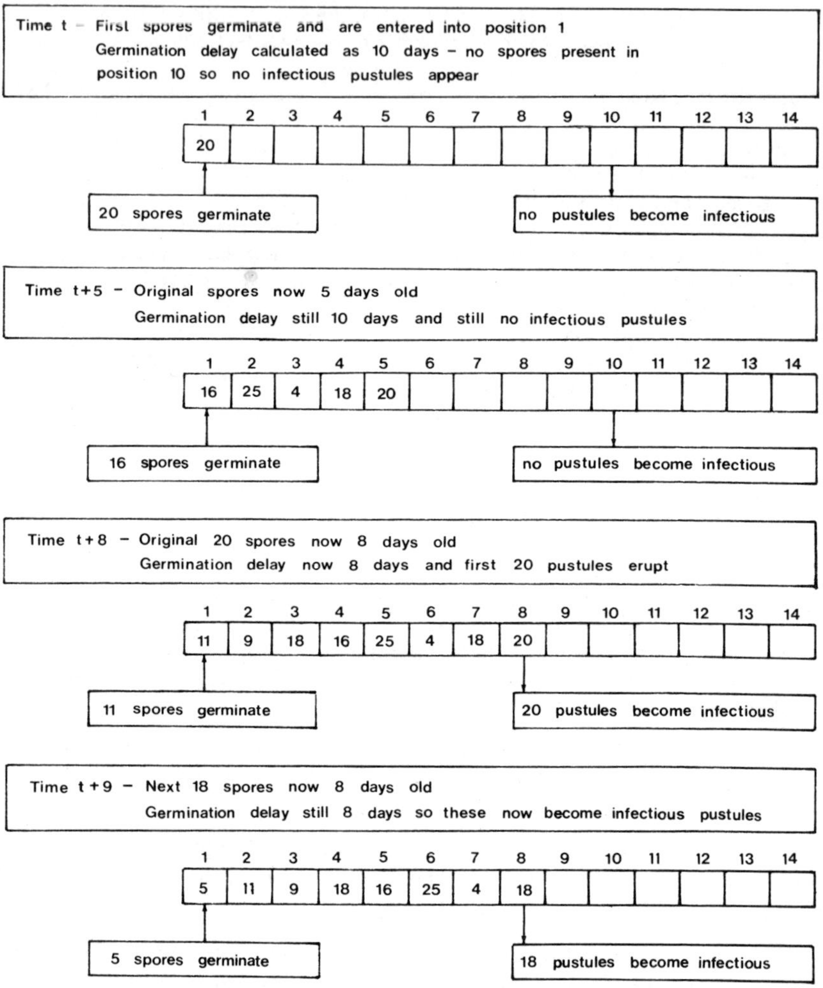

Fig. 2.7. Operation of event-stepping module.

If a time-stepping mode is selected for the model, the size of the time interval must then be defined. Time, of course, moves continuously forward in the real system; the intervals of seconds, minutes, days, and so on, are merely counting devices. In a model established on a digital computer, however, it is not possible to move time in a continuum since such machines are sequential processors. The time interval selected is directly related to the level of detail required in the model. A detailed biological model of ruminant digestion might require a time interval of minutes or even seconds. A similar model for evaluating alternative feeding policies on a farm might need a time interval of a day or a month. Selection of the basic time interval for the model is a complex problem which will be mentioned again in later chapters. In many biological situations, however, models may need to contain both time-stepping and event-stepping modules.

Stage 4a

Assuming adoption of the time-stepping notation we can begin to establish the equation form that can represent the symbolic model. We will be concerned with the change in a 'level' between points in time.

Return to the example of the children playing in the filling bath tub. Here the basic equation will be

$$\text{water level}_t = \text{water level}_{(t-1)} + \text{any inflow} - \text{any outflow}$$

The inflow and outflow will be the rates occurring between time $(t-1)$ and time t. In more formal terms the difference equation becomes

$$LW_t = LW_{(t-1)} + (WIN_{(t-1)\rightarrow t} - WOUT_{(t-1)\rightarrow t})$$

where

LW_t is the level of water at present time;
$LW_{(t-1)}$ is the level of water at previous time;
$WIN_{(t-1)\rightarrow t}$ is the inflow of water between $(t-1)$ and t;
$WOUT_{(t-1)\rightarrow t}$ is the outflow of water between $(t-1)$ and t.

For the sake of this discussion assume that the rate of inflow of water depends on how many other taps are turned on in the household at the same time; taps for the washing machine are time-of-day dependent, taps for the garden hose dependent on past rainfall, taps for the car-washing hose are also dependent on past rainfall since this determines the amount of dust on the roads, and so on. Further assume that the outflow of water is a direct function of the number of children in the tub at one time. This will change over the bath-time period as children come in from play (time and weather dependent) and as children leave the bath for bed (time dependent). Suppose that the bath-time period is a variable number of minutes (M) and that we wish to represent change every second over that period (recall the time interval decision is complex so at this point in the text no attempt to justify this interval will be made).

The symbolic model will continue from the simple difference equation above to represent the factors affecting the rates of inflow and outflow of water.

$WIN_{(t-1)\rightarrow t} = f(NTAP)$ —water inflow dependent on the number of other taps turned on;

$NTAP = f(t, RAIN)$ —number of other taps turned on dependent on time and past rainfall;

where
$NTAP$ is the number of taps turned on
and
$RAIN$ is the amount of rainfall in the past (calendar month).

$WOUT_{(t-1)\rightarrow t} = f(CHILD)$ —water splashed out dependent on the number of children in the tub;

$CHILD = f(CHILDIN, CHILDOUT)$—number of children dependent on children coming in and children getting out of the tub;

$CHILDIN = f(t, PRAIN)$ —number of children getting in dependent on present rainfall and time;

$CHILDOUT = f(t)$ —number of children getting out is time dependent;

where
$CHILDIN$ is the number of children going into the tub.
$CHILDOUT$ is the number of children leaving the tub.
$PRAIN$ is today's rainfall.
Both $RAIN$ and $PRAIN$ are stochastic functions of time of the year so

$RAIN = Sf(t)$ —past rainfall is a stochastic time-dependent variable

$PRAIN = Sf(t)$ —present rainfall is a stochastic time-dependent variable.

In this way the non-quantified symbolic model is assembled in a progressive way beginning with the basic difference equations concerned with the levels in the model and working through the rates which influence the levels to the factors influencing these rates. The link to the *Industrial Dynamics* format is clear and underlines the fundamental value of its conceptual basis.

Using the FORTRAN language we will now pursue the basic symbolic model a step further but without, at this stage, considering the stochastic input of rainfall.

The essential work now is to establish the clock for the model within the predefined time interval of 1 s and the total time for the bath-time period of M min. The performance parameters for the system are the total amount of water inflow to the tub and the level of water in the tub at the end of the period. Within this period we will wish to monitor the amount of water in the tub every second and every minute and the number of children in the tub at 1-min intervals. The clock we must develop, therefore, must tick through 60 s in each minute up to M min and each second must be specified uniquely so that we can record at the end of each second how much water is in the tub.

Let M be the number of minutes for the bath period.

IMIN (1, 2, . . . , M) index the minutes;
ISEC (1, 2, . . . , 60) index the seconds within each minute.

Specifying M, the clock can be reconstructed by two nested DO-LOOPS as follows:

```
            READ (5, 10) M
            DO 100 IMIN = 1, M
            (calculations on numbers of children are in this loop)
            DO 101 ISEC = 1, 60
            (calculations on water level are in this loop)
101         CONTINUE
100         CONTINUE
            STOP
            END
```

The nested DO-LOOPS of the model-clock not only form the framework for the computer version of the symbolic model but also allow us to set out the dynamics of the basic symbolic model. Within each pass of the DO-LOOP for seconds, the difference equation is called upon to update the level of water in the tub. To do this the rates of inflow and outflow also must be calculated for each pass. Within each pass of the minute DO-LOOP we simply arrange to monitor the level of water in the tub and the number of children present at the end of each minute.

With this framework we are well on the way to computer implementation but generally (particularly for more complex models) a formal diagram is drawn up prior to the final computer-programming work. This stage is described below.

Stage 4b

Up to this point the structure of diagrammatic models has been informal, in which the symbols and interconnections used are of a form most suited to the individual model-builder. In setting up a formal flow diagram, however, it is recommended that a standard format such as that of the International Organisation for Standardisation Draft Recommendation on Flowchart Symbols for Information Processing is applied. Standard templates for the symbols involved can be purchased readily; in Fig. 2.8 the five most common symbols are illustrated so that we can proceed to the formal flow chart for the bath-tub model.

Using the model-clock in FORTRAN format and the basic symbolic model previously established, we can set up the flow chart for the bath-tub model as in Fig. 2.9. The flow chart is established around the two nested DO-LOOPS and within each loop the various calculations are set in 'Process' symbols.

The disadvantage with flow charting for computer simulation models is

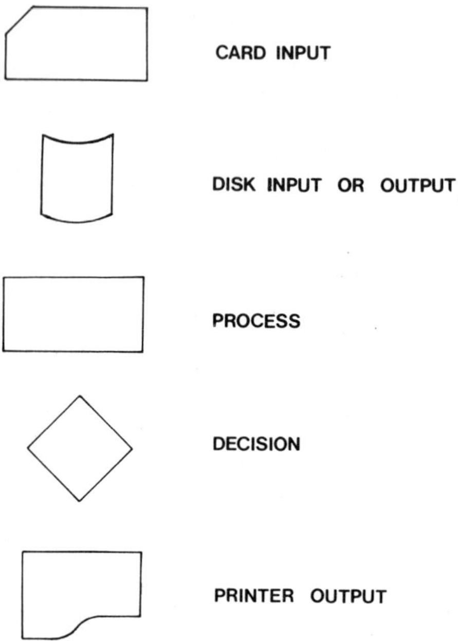

Fig. 2.8. Some commonly used flow-chart symbols. (After IBM flow-charting template.)

that drawing up a flow chart for a complex system can be a long and tedious business. Frequently the flow chart does not get drawn up until late in the model-building process. Even then, because of the difficulty of modifying flow charts, these charts are frequently not kept up to date.

An alternative is to use structure diagrams which are increasing in popularity among model-builders. The structure diagram in concept falls between the block diagram and the flow chart. The human mind prefers to consider problems as hierarchical structures whereas the computer works in a sequential logical fashion. The structure diagram is an attempt at visually displaying a hierarchical structure in a format which can be readily coded into sequential instructions. It is worth emphasising that this notation accords well with the systems concepts outlined in Chapter 1.

Three basic program structures are

1. a simple sequence of steps;
2. a repeated sequence of steps (i.e. a loop);
3. a choice between two or more steps.

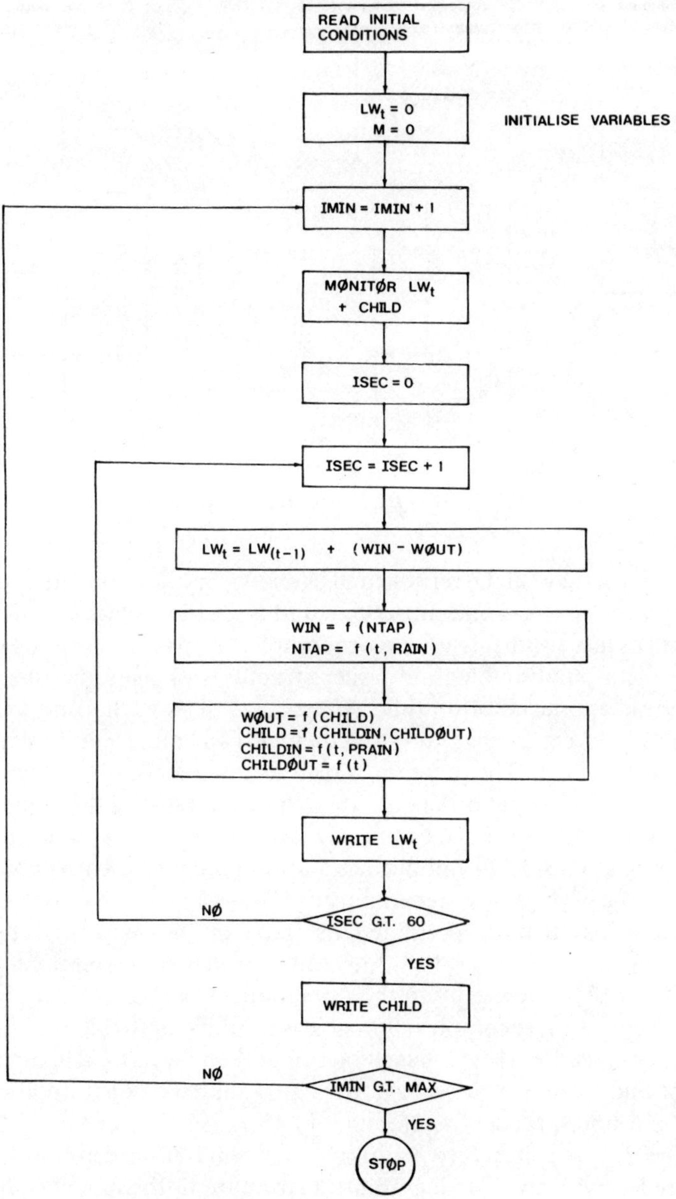

Fig. 2.9. Flow chart for bath-tub system.

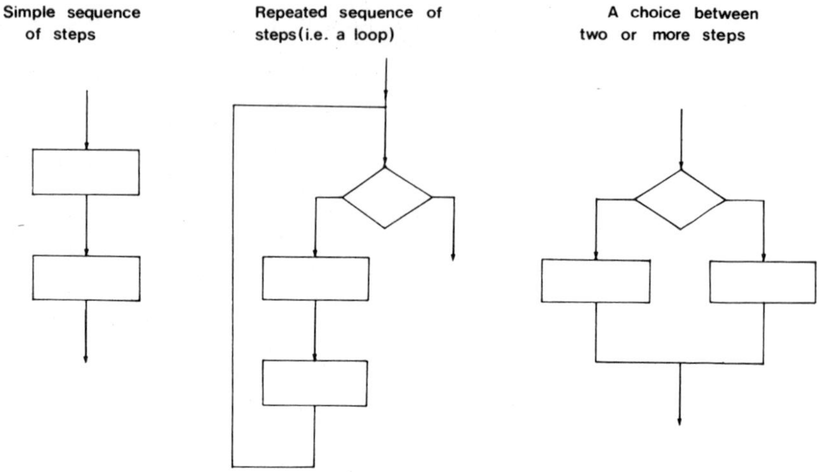

Fig. 2.10. *Basic program structures.*

In chart form they can be represented as shown in Fig. 2.10. Note that each of these structures has one entry point and one exit so that each one can be regarded as just another (rather more complex) step in the program and can be used as a building block in larger structures of the same three types. Structure-diagram notation differs from that of flow charting and some examples of structure diagrams together with an equivalent flow chart are shown in Fig. 2.11. The diagram is read as a sequence of steps which are executed in left-to-right order and when the last is finished the sequence as a whole is considered finished. A structure diagram is well suited to developing alongside a symbolic simulation model as it corresponds very closely to the process of systems analysis. The system is broken down into a series of subsystems (represented as steps in the diagram) and these subsystems are in turn broken down into smaller subsystems (steps on a lower sequence), eventually to their component parts. Particularly as the model becomes more complex, flow charts are difficult to follow and require the ability to read in a looping fashion rather than the natural manner of left to right and from top to bottom. Structure diagrams offer an alternative and often more practical approach.

Figure 2.12 is a structure diagram of the children-in-the-bath situation described previously. The hierarchical structure of the model is apparent from the diagram and the assumptions which have been made with regard to factors such as the sequencing of events are clearly laid out. For example,

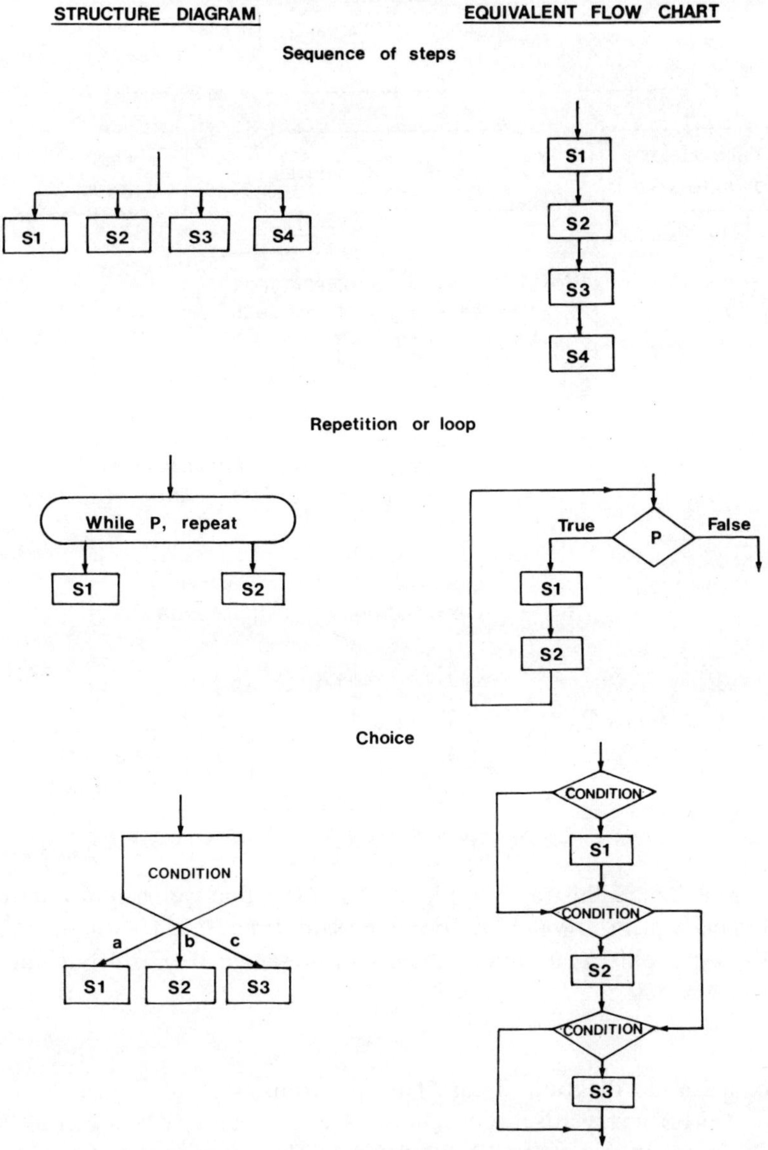

Fig. 2.11. *Structure-diagram notation.*

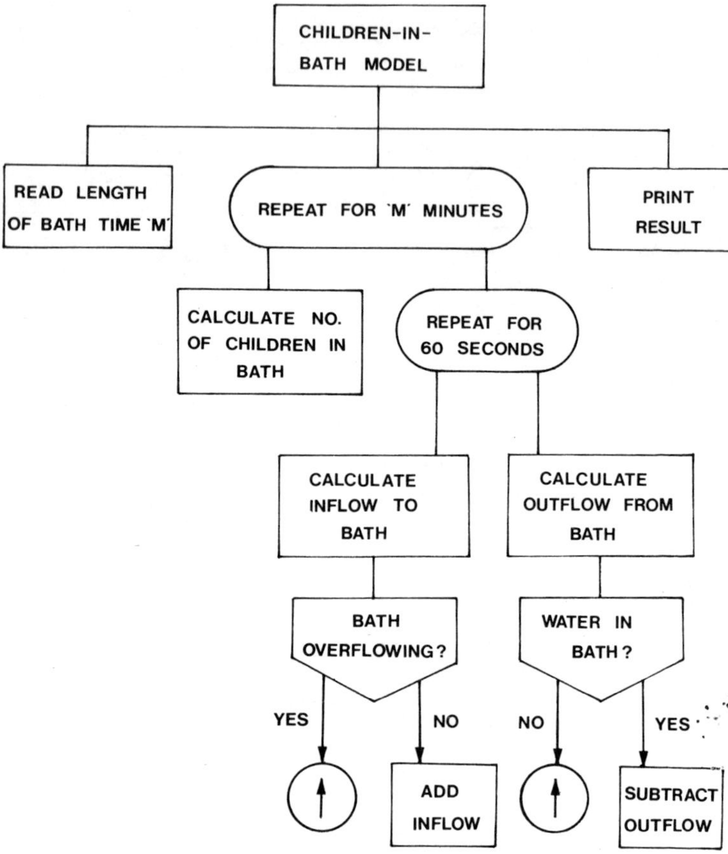

Fig. 2.12. *Structure diagram for bath-time system.*

the simple model illustrated in Fig. 2.12 assumes that water inflow in time t takes place prior to water outflow in the same time period—an important and questionable assumption which was present in Fig. 2.9 but not so clearly obvious.

Stage 5

Once a formal diagram is complete, the translation of the model into computer form is relatively straightforward. Problems associated with the choice of a suitable computer language will be considered in Chapter 3. Here let it suffice to say that although the range of languages is large, in practice the modeller will generally have a greatly restricted choice.

Obviously facility with the chosen computer programming language is essential by at least one member of the modelling team. Preferably all concerned with the construction of a simulation model should have some knowledge of computer programming to aid communication within the group. At this stage it is worth noting that there are procedures which can reduce the likelihood of errors ('bugs') being built into the computer model and to detect any such bugs which may inadvertently be included. (Programmers are human and in any reasonably complex model, bugs can be expected.) These procedures will be discussed in detail in Chapter 3.

Alongside the development of the computer model it is most important to fully document the process of computer implementation as it is carried out. A complete record of the computer-programming work should be kept so it is clear exactly what has been done as the translation into computer code proceeds. The issue of documentation will be addressed in subsequent sections and we will not labour the point here. Although documentation may appear to be tedious, invariably it is a great time saver. Many models will be developed over months of work and re-evaluation of earlier developed parts of the model will invariably be required. Such reappraisal can be a frustrating business if proper documentation has not been maintained, particularly where a team is involved:

At the completion of Stage 5, we should have a basic model which operates in a reasonably acceptable fashion and which can be further developed, formally tested and evaluated and eventually put to some use. However, before moving on to consider these issues, we will backtrack a little and consider in rather more detail some of the particular problems associated with translating the model-concept into an actual working computer model.

Summary

In this chapter, we have gone step by step through the process of looking at a system, analysing it into its component parts, reassembling these components in a symbolic form and finally constructing a working computer model. This, hopefully, has enabled the reader to gain an overall picture of this most important stage of model-building. Not all the detail necessary to implement the various stages described has been provided in this chapter but this will be rectified in the following chapters. One of the main emphases of this chapter has been to illustrate how time is incorporated into simulation models. The various ways of treating time have been discussed and illustrated with examples. We have also considered in some detail the problems associated with obtaining satisfactory data for modelling purposes and the methods whereby the model-builder may get round difficulties of missing or inadequate data.

Appendix

This appendix describes in some detail models of the two systems that provide the main examples for this text, i.e. the barley-leaf-rust agroecosystem and the pig-production system. The two systems concerned, and our objectives in modelling this, have already been described in the appendix to Chapter 1. Here we illustrate how the modelling process commences in order to achieve the stated objectives. Structure diagrams and flow charts for the models will not be presented but the reader is invited to produce for himself diagrammatic representations of these models at this stage.

BARLEY-LEAF-RUST MODEL

The rust model is conceived as two main segments—one which simulates the disease epidemic and the other which estimates the yield loss due to the simulated epidemic. The main biological and physical components in the epidemic segment are illustrated in Fig. A2.1 (Teng, Blackie & Close, 1977) and are an infectious period (which is the time between pustule eruption and pustule death), spore production, liberation, survival, deposition, germination, penetration of host tissue by germ tube and, finally, a latent period (which is the time between penetration and pustule eruption). These components were established from field and laboratory studies of the leaf-rust pathogen. Having defined the life-cycle of leaf-rust fungus, the next task was to estimate the loss in yield in a barley crop affected by the disease. A measurement of the progress of the disease had to be established and, following the work of James (1971), the area of infected tissue expressed as a percentage of total tissue area was used. This concept is expressed in the term 'percentage severity'. Yield loss was then estimated from a function relating percentage yield reduction to the percentage severity of disease on leaf 1 (flag) and leaf 2 of tillers at certain crop growth stages as defined within the decimal code of Zadoks, Chang and Konzak (1974). The relationship between disease progress and projected crop loss was derived from field experiments at the following crop growth stages; early boot, beginning of flowering and milky dough on the Feekes' scale.

The model has been established to operate on a daily time interval ($\partial t = 1$ day) and the starting period of any simulation is the day on which disease is first observed in the field. Data input to the model include percentage severity of initial observation of infection, number of days to be simulated and weather data.

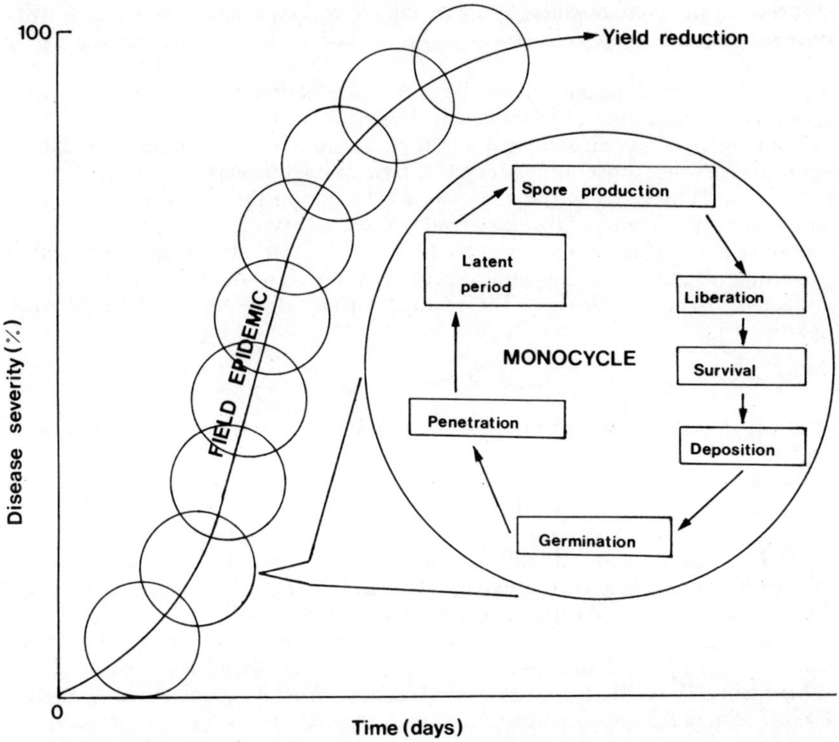

Fig. A2.1. The rust epidemic and components of its monocycle.

The liberation of spores by an infectious pustule rises to a peak early in the infectious period and then declines gradually through the life of the pustule. In the model, therefore, the potential number of spores produced on any day is a function of the age distribution of pustules and the number of infectious pustules. The modelling of several events in the fungus life-cycle involved stochastic elements† and the model was designed so that such elements could be either stochastic or deterministic depending on the requirement for a given run of the model. In a deterministic run, the actual number of spores liberated equal the specified potential, while in a stochastic run, actual spore number liberated is the result of multiplying the number of spores produced by a daily correction factor. This correction factor was drawn from a uniform random-number generator (0 to 1) for each day of the simulation. The deterministic or stochastic nature of the survival and deposition phases are

† A full discussion of stochastic elements will be presented in Chapter 4. For now it is sufficient to say that a deterministic event will always result in the same output from a given input. A stochastic event will produce a variable output dependent on some defined probability density function from the same given input.

modelled in the same manner. Spore germination is regarded as a binary activity since it is dependent on leaf wetness. (A binary activity is one which can have only two values: in this case germination either does or does not occur.) If dew is present, then all the spores deposited on that day germinate; if dew is absent, then no spores germinate. A definition of when dew is present is clearly involved.

The number of germinated spores that penetrate to form pustules is dependent upon the rust-race–host-cultivar combination. While the expected success rate of penetration is included in the model as a 'penetration ratio', this relationship is modified by the amount of disease already present. Rusts are obligate parasites and can only successfully infect healthy tissue. Therefore the model includes an important negative feedback with respect to penetration success as the amount of uninfected tissue becomes less. This relationship corresponds to the mathematical notation $(b - y)$ in Verhulst's logistic equation

$$dy/dt = ay(b - y)$$

where dy/dt is the rate of growth, a is a rate parameter, y is the amount of growth at time t and b is the maximum attainable growth.

Output from the leaf-rust model includes a daily estimate of disease severity and an estimate of percentage yield at the end of the simulation run. Disease severity may be plotted to assist visual appraisal of the results of each run.

Model-building involved two main phases. An initial system model (Phase 1 model), which represented the epidemic caused by the urediniospore stage of rust diseases, was constructed after a general literature review of the rusts. Preliminary experimentation with this model enabled location of critical control points in the system and suggested the degree of emphasis to be placed on various system components (Teng, Blackie & Close, 1977) in subsequent development of the model. Empirical data for the structure of the detailed system model (Phase 2 model) was derived experimentally from controlled-environment and field experiments. This involved determination of the effects of weather parameters on development of the fungus on barley, the effect of the fungus on barley yield and the effects of fungicide on the progress of the disease epidemic. The detailed system model was assembled on a modular basis, with component subsystems linked by a main executive routine responsible for the time-keeping and event advancement (Dent, 1974). This model will be incorporated into an agricultural-information system along the lines outlined by Blackie (1976). Farmers adopting the system will be able to use it to perform cost-benefit analyses on the economics of using fungicides to control barley leaf rust in the field during the growing season. The system will provide the individual farmer with a rational basis on which to base fungicidal applications and should play a positive role in reducing the unnecessary use of pesticides.

PIG-HERD MANAGEMENT MODEL

The pig-herd model is conceived as two main subsystems, a breeding unit and a feeding unit. Each area in turn is subdivided into three main modules:

1. modules for generating the flows of pigs through the system;
2. feeding modules to predict food requirements;
3. modules to calculate the cash flows produced by 1. and 2. above.

The simulation of the enterprise is performed by those routines which forecast the flows of pigs through the system. For the purposes to which the model is directed it is not necessary to identify pigs as individual animals. Instead, they are grouped according to their reproductive stage for breeding sows and liveweight for fattening pigs. For example, fattening pigs are grouped into fairly narrow weight ranges and sows are classified as in-pig, suckling or awaiting service; this latter classification is further subdivided to group together sows which will complete a stage in their breeding cycle in a particular week.

Figure A2.2 illustrates the basic mechanism used to forecast pig movements within the breeding unit. The minimum time step in the model is 1 week. It is assumed that all sows completing the suckling activity in any week are weaned in that week. Following this, cull sows are withdrawn and any remaining sows are entered into the anoestrus activity† which is assumed to be of 1 week duration. The model is constructed so that as sows complete their first anoestrus they, together with any replacement gilts at the correct reproductive stage, are served. Since in the real system a proportion of sows will fail to hold to service and will enter a second anoestrus phase (second and third anoestrus are assumed 3 weeks in duration) a mechanism must be developed to permit the proportion of sows failing service to be a function of first anoestrus length, weaning policy and recorded farrowing index (average number of farrowings per sow per year). Sows that have been successfully served are assumed to be held in the 'in-pig' activity for a gestation period of 17 weeks. At the end of this time they are transferred to the start of the suckling activity. At the same time, piglet births are calculated according to expected number of piglets per litter specified in the input data, and piglets enter the start of the piglet activity. They are held in this activity until the age at which they are weaned and either sold or transferred over to the fattening unit.

Replacement gilts come from two sources: gilts retained from pigs bred on the farm and gilts bought in from outside the system. Gilts retained on the farm are transferred back from the fattening unit to the breeding unit at the requisite age. Gilts are only bought if insufficient numbers are available for transfer to compensate for sows culled or if no requirement for retaining breeding gilts is stated in the input data.

Culling and replacement of breeding stock is based on one of two alternative methods of data input. For a policy which simply intends to replace a given proportion of the herd each year, the percentage of sows or boars due for replacement each year must be specified. In this case, the model is organised to use these data to estimate the number of animals to be culled each week, to cull the requisite number of animals and to introduce sufficient replacements to keep herd size constant. An alternative is provided for the situation where the herd size is to be altered. Now, the numbers and intended time of purchase of relevant stock must be stated and the model will operate according to these instructions. In all cases, culling requirements are satisfied first from infertile sows (as defined above), and if these are insufficient from sows that have completed suckling.

In the model the breeding unit and the fattening unit can either operate jointly or independently. Figure A2.3 illustrates the operation of the fattening unit. Pigs

† An explanation of the basis for determining the number of cull sows and replacements follows below.

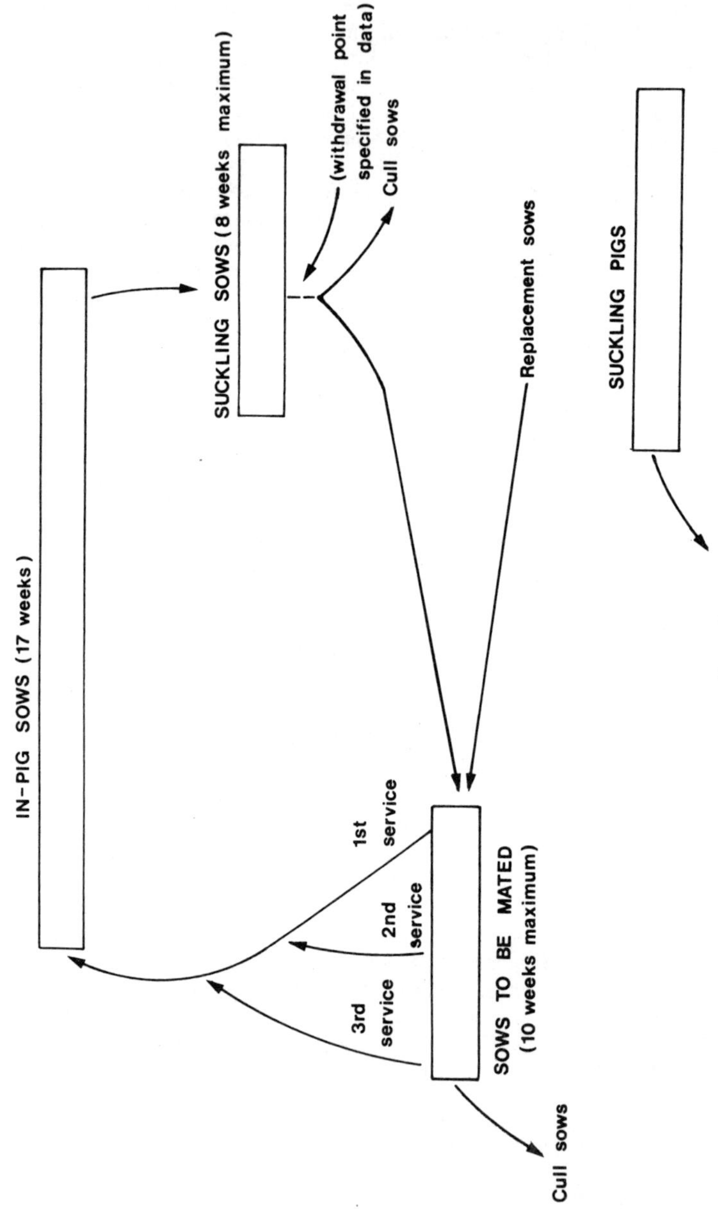

Fig. A2.2. Dynamics in the breeding unit.

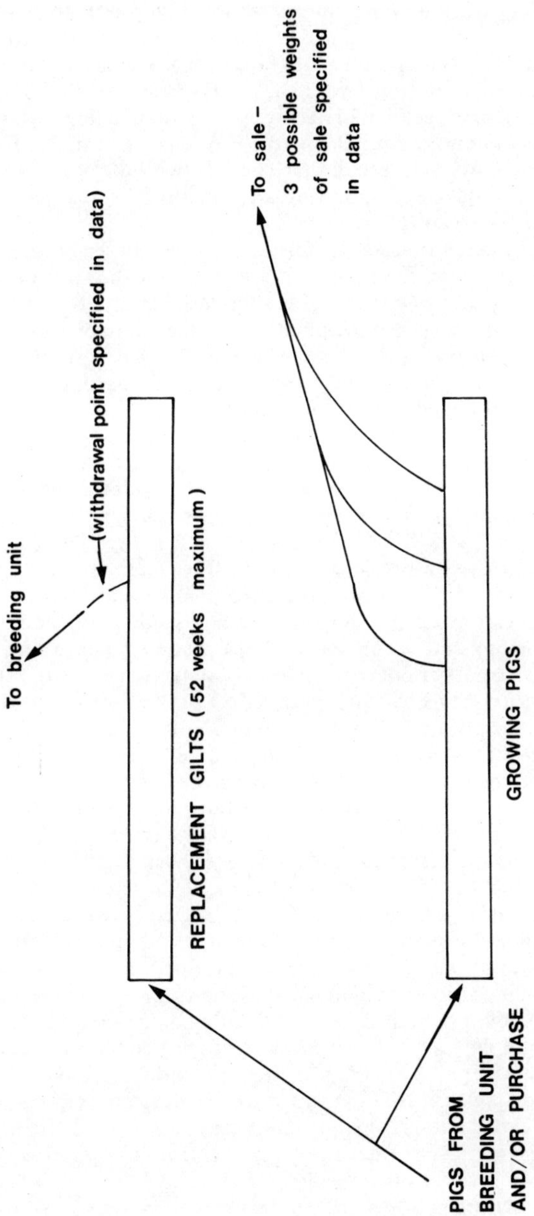

Fig. A2.3. Dynamics in the fattening unit.

entering the unit come from the breeding unit or as a result of a specified buying policy. Clearly, where the fattening unit is operated independently, a buying policy must be stated. If gilts are to be retained for eventual transfer back to the breeding unit, a proportion of pigs entering the fattening unit are separated for this purpose. Gilts of the requisite age are transferred back to the breeding unit. The proportion retained is a function of the estimated weekly wastage of breeding sows (based on the stated annual sow-culling rate for the herd). Provision is available for additional gilts to be withdrawn from the regular fattening activity, if pigs of an appropriate age are available, should the replacement needs in any period exceed the numbers available in the replacement activity.

After gilt replacement requirements for the week, if any, have been fulfilled, the remaining pigs entering the fattening unit are placed at the start of the fattening activity. This activity allows pigs to grow at a variable rate dependent upon the number of weeks to reach the top weight at which they may be sold. Variability in growth rate is imposed by assuming that the age at which pigs reach the top sale weight is normally distributed. The appropriate standard deviation for this distribution can be obtained either from data for a particular policy or a model-default value. Growth of pigs in the fattening activity is assumed to be a function of the ration fed (three sequential rations are allowed), the feed conversion on that ration, the time over which the ration is fed and the standard deviation of days to slaughter weight. As pigs achieve sale weight, they are sold in accordance with the selling policies provided as data.

The feeding and cash-flow routines are quite simple mathematical appendages to the model and do not, in their own right, perform any event-forecasting function. The feeding routines for the breeding and feeding sections of the model operate differently. For the breeding sector, each type of pig defined in the input data—for example, in-pig sows, gilts or boars—can be fed a different ration and suckling sows may also be fed according to litter size. Model-estimates of food use in any particular week is the product of the total numbers of each type of pig in that week and the feed rate for that pig class specified in tha data. The cash-flow routines apply monetary values to the various inputs and outputs to and from the model. Some of these inputs and outputs have a variable cash value—for example, failing to meet a contract may result in a lower price per pig. Allowance is made within the model for such cases. Normally, the cash values of inputs are constant during the whole of the forecast. It is possible, however, to include an inflationary trend into prices and also to supply separate sale prices for fat pigs for each period of the forecast. Data input to the model include the asset structure of the unit and annual charges including labour costs, feeding policies in the various production stages of the unit and estimated performance rates (including carcase assessment), and replacement and sales policies for breeding and fattening stock. Other data required are performance standards (for example, mortality at various stages or piglet production per sow), information on the current state of the system (including numbers and estimated weights of fattening pigs and the numbers of sows at different stages of the breeding cycle), and price expectations for inputs and outputs.

Output from the model is a forecast of herd behaviour in physical and financial terms. This forecast is either on a weekly or a 4-weekly basis and is for the next 12 months. The information provided in the forecast enables a manager to compare the results from specified strategies prior to selection of any particular one.

The model-building phase was relatively straightforward. The model, largely as described above, was constructed and validated against data collected on farms during the model-building phase. The model was then tested in the field and various features added or deleted. For example, at one stage the inclusion of a detailed price-forecasting module was considered since price variability is one of the major constraints in pig production. However, field experience showed that farmers using the model preferred to use their own price expectations. The price-forecasting module was, therefore, dropped but other modules—such as one to mimic particular contract-setting arrangements—were added as need indicated.

Workshop

1. Draw a block diagram representing the pig system described.
2. Use *Industrial Dynamics* symbols to establish a diagram of the barley-leaf-rust model.

References

Blackie, M. J. (1976). Management information systems for the individual farm firm. *Agr. Systems*, **1**, 23–36.

Dent, J. B. (1974). *Application of Systems Concepts and Simulation in Agriculture*. School Agriculture, University of Aberdeen, Misc. Pub.

Forrester, J. W. (1968). *Industrial Dynamics*. Cambridge, Mass.: MIT Press.

James, W. C. (1971). An illustrated series of assessment keys for plant diseases, their preparation and usage. *Can. Plant Disease Survey*, **51**, 39–65.

Shannon, R. E. (1975). *Systems Simulation: The Art and the Science*. Englewood Cliffs: Prentice-Hall Inc.

Teng, P. S., Blackie, M. J. & Close, R. C. (1977). A simulation analysis of crop yield loss due to rust disease. I. Model development and experimentation. *Agr. Systems*, **2**, 189–98.

Zadoks, J. C., Chang, T. T. & Konzak, C. F. (1974). A decimal code for the growth stage of cereals. EUCARPIA Bull. No. 7.

3

Computer Considerations

The time has now arrived in the model-building process to consider the implementation of the model on the computer. Inevitably the discussion moves towards the selection of an appropriate computer language. In this book the language FORTRAN is used in the text examples because it is a widely understood code. While this is a consideration in selecting a language for a particular modelling exercise it is by no means the only or the most compelling basis for selection. Once this decision is made the translation from the flow diagram to the selected language must be made. This is a time when the fallibility of the human mind can become more than a simple irritant as mistakes and errors inevitably creep into the translation. This chapter has two aims:

1. to outline the arguments in the determination of a suitable computer language for a specific modelling project;
2. to provide guidelines for translating the model into the selected language in such a way as to limit the number of errors.

Choosing a Language

A large number of different computer languages suitable for simulation purposes have been written, and numerous articles have already described and compared these in considerable detail. A number of valuable statements on this topic are already available but invariably most of these descriptions have been written from the point of view of the specialist computer programmer with the result that others find them confusing rather than helpful. This section will consider computer languages in a largely non-technical way, introducing some of the basic ideas behind their design before discussing factors affecting the selection of suitable languages.

THE DESIGN OF COMPUTER LANGUAGES

The computer operates by using a series of single-operation instructions which are supplied to it in the form of a program. While ultimately all programs written for a computer are broken down into these single-operation steps, most computer users make use of the convenient and efficient 'high-level' computer languages which are available on modern computers. These high-level languages use simple English-like verbs and standard mathematical notation to represent a series of single-operation instructions. For example, the input of a single data item through the card reader requires that the computer be instructed, among other things, about what form the data item is in (alphabetic or numeric, for instance), where the data item is to be found (on the next computer card), in what position in memory that data item should be stored and finally where, in computer memory, to find the next step in the program to be completed. All this can be replaced by single high-level instructions as in the FORTRAN example below.

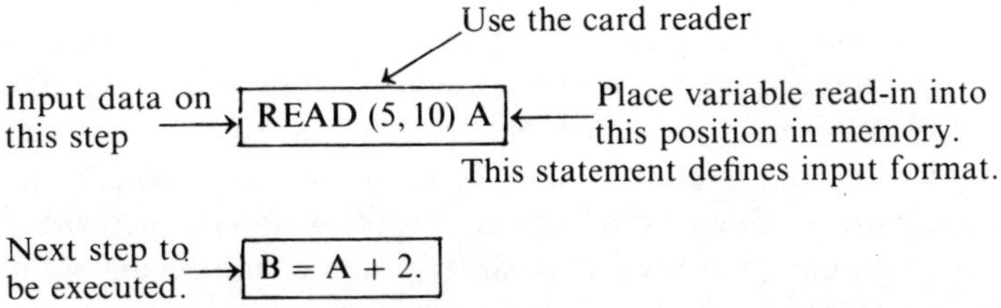

The translation from a high-level language into the eventual series of single-step instructions is referred to in computer jargon as 'compilation'. A compiler is a specialist computer program which converts a high-level language into single-step instructions. Each high-level language has its own compiler which requires that the programmer follow specified language rules. These rules are referred to as the 'syntax' of the language. Before any program written in a high-level language can be run by a computer, the compiler checks the program for syntax and if the program fails this check, the computer will reject the program.

The two statements below illustrate how syntax varies between languages. Both statements result in the variable 'A' being assigned the value 20. The first statement, however, is in FORTRAN and the second in ALGOL and the reader will notice several small differences between the two statements.

A = 20.	FORTRAN assignment statement
A : = 20.;	ALGOL assignment statement

The next example shows a FORTRAN statement which has incorrect syntax. The compiler assumes that brackets always occur in pairs and, in this example, the right-hand bracket is missing.

A = (20 * B/C + 4.

It cannot tell whether the programmer meant

A = (20 * B)/C + 4. or
A = (20 * B/C) + 4.

and therefore rejects the statement as being insufficiently well-defined for execution. Had the programmer left out both brackets, certain default rules for execution of this statement would have been followed.

High-level languages make programming easier to understand (both for the programmer and the non-programmer), and make the task of programming simpler and less tedious. While high-level languages are designed to make programming easier, there are important considerations to be taken into account in their design which can influence their value and their power. An analogy to natural languages will perhaps help a little. As the vocabulary used becomes larger, the phraseology more complicated and the construction more flexible, then the quality of the mind required to understand and interpret the meaning of the words has to be correspondingly greater. Similarly, any increase in the 'power' and flexibility of a computer language requires a disproportionate increase in the complexity of the compiler needed. A larger compiler will occupy more of the computer memory and may be more expensive to use and, inevitably, will have a greater likelihood of containing 'bugs' of its own.

A more powerful language may also be more difficult to learn sufficiently well to make full use of its potential capabilities and may lead the unskilled and unwary modeller into unintended errors of logic. Some compromise, therefore, has to be made between the power of the language, the computer facilities available and the knowledge and experience of the programmer involved. Broadly, computer languages may be grouped into two classes— general purpose and special purpose— and the main features of each class are described briefly below.

1. *General-purpose languages*

Into this category fall such languages of interest to the simulation modeller as FORTRAN, ALGOL and PL/1. Because of their generality, they may be used to construct simulation models of any type of system. However, the modeller has to develop his own input–output routines, set up his own time clock and switches within the model and write his own special-purpose routines such as normal pseudorandom-number generators. Modelling using these languages requires programming expertise and a reasonable knowledge of the computer and its associated systems.

2. *Special-purpose simulation languages*
These languages contain specialised facilities which are convenient for modelling particular types of situation. In the main, the objective is to make it easier for the non-programmer to write his own model without the need of specialist-programmer assistance. Automatic time-keeping routines and sophisticated output facilities are often incorporated in such languages. Some of these languages are supersets of some general-purpose language such as FORTRAN (CSMP is an example) whereas others are self-contained such as GPSS. Most of these languages were originally constructed to satisfy the requirements of specific problems and they, therefore, differ in the type and range of their possible application. If the modeller's requirements do not quite correspond with those provided by the language, then there may be considerable problems in adequately representing the real system. The language may then begin to dictate the structure and operation of the model rather than the real system.

CHOICE OF GENERAL-PURPOSE OR SPECIAL-PURPOSE LANGUAGE
There are four main considerations in deciding whether to use a special-purpose or a general-purpose language.

1. *Availability*
Only use a language that is readily available and fully supported by a handy computer centre. Languages are software which are supported either by the computer manufacturer, the computer-centre systems staff or by the language developer. It is not generally advisable to use a language which does not have a compiler readily available on the installation being used. Advice should be sought from the computer centre as to availability of, and support for, any language being considered. The would-be modeller will quickly become discouraged if the language compiler is erratic, slow or expensive to use.

2. *Existing programming skills*
For many modelling exercises, the extended general-purpose language sets available on modern computers combined with selected use of library software can provide many of the facilities of a special-purpose language.

The modeller using a general-purpose language will have to develop his own input–output routines, data-checking facilities and time counters but this must be balanced against the not-inconsiderable time required to learn adequately a special-purpose language.

3. *Machine independence*
Where a model is to be used at several different centres, the range of languages (and facilities within a given language) will invariably be significantly constrained. It is clearly imperative to use a language which is supported at all centres.

4. *Cost*
Many special-purpose languages invoke both a high compilation and run-time cost. The high cost of compilation may be offset against a possible saving in time for model-development. However, where a model is to be repeatedly used, a general-purpose language may, in this regard, offer considerable long-term savings.

In general, the bias of the authors is towards either a general-purpose language or a language such as CSMP which is a superset of a general-purpose language. This is largely because experience indicates that these languages are generally well-maintained and supported by computer-centre staff and that models using these languages are more easily transferred to other centres.

SELECTION OF A SPECIAL-PURPOSE LANGUAGE
If the model-builder is considering using a special-purpose language then the problem arises as to which to choose. Charlton (1971) and Shannon (1975) both provide excellent reviews of the various languages and the latter author sets out a decision chart for selecting among them. Many of the available simulation languages are designed for military or industrial processes and may not be suitable for biological or specifically agricultural applications. As a broad guide to selecting a suitable language, it is possible to divide the systems likely to be of interest to the biologist or agriculturalist into two main categories (Teichroew & Lubin, 1966). Charlton (1971) cites examples for these two main categories of system: the growth of a plant illustrates the system which is continuously changing through time while the grain-harvesting system represents the second type of system concerned with a sequence of operations each occurring at a discrete point in time.

In the plant-growth system a number of interdependent, continuously-changing relationships can be envisaged as being responsible for the development of the whole plant through time (water balance, photosynthesis, respiration, nutrient balance). Not only are these relationships interlinked but they also exhibit marked feedback conditions—for example, the rate of photosynthesis influences the rate of tissue development which in turn influences the leaf area which influences the rate of photosynthesis. Rates of change for processes can be integrated over time to give levels for major plant components (status variables in the simulation model). The key characteristic for languages which seek to represent such systems is the capacity to integrate sets of differential (rate) equations. A number of languages which may be described as 'integrating' in nature are available at many computer centres. Examples of suitable integrating languages are CSMP and DYNAMO. In a sense these two languages are at the opposite extremes for integrating languages: CSMP (particularly in its CSMP/360 form) has a wide choice of approximate integration procedures which can be selected by the modeller depending on the nature of his model. This can be a most attractive characteristic in modelling biological systems with their complex interacting continuous relationships. DYNAMO on the other hand can only cope with simple rectangular integrations of a well-behaved type. This is illustrated in the DYNAMO level type of equation.

$$LEVEL_t = LEVEL_{t-1} + DT\ (RATEIN \pm RATEOUT)$$

Such an integration can, of course, be included in CSMP as one of the options.

The grain-harvesting system is not at all concerned with integrating rates of change over time because the whole process is essentially a sequential one: the ripe crop is harvested with a combine harvester, the grain is transported in sacks or in bulk trucks to the farmstead where it may be dried and cleaned before transfer to storage bins. Reasons for studying this type of system will be to alleviate bottlenecks in the process by investigating factors causing delay—harvesting capacity, grain transport, unloading capacity, etc. Here it is only necessary for the model of the system to advance through time, recording the occurrence of events and executing decisions in the appropriate chronological order. Examples of specialised languages of a non-integrating type are GPSS and SIMSCRIPT. Of the two, SIMSCRIPT which permits the inclusion of FORTRAN statements between its special statements is the more flexible and may be more generally preferred in representing sequential systems.

For the sake of this discussion we have made a binary classification of systems into continuous and discrete and discussed special-purpose simulation languages as integrating and non-integrating in relation to this classification. Most systems cannot so readily be pigeon-holed in that they contain continuous elements which need to be described by differential equations as well as discrete elements which involve sequential queueing processes. The strength of special-purpose languages which can also call on the FORTRAN library is that FORTRAN statements can be used to describe the alternative elements (either the continuous or the discrete) while the special-purpose library can give more sophisticated aid in representing the prime elements. So in using CSMP, special-purpose integration is provided, with FORTRAN able to cope with any discrete processes, and SIMSCRIPT can deal with sequential and queueing aspects while FORTRAN provides a facility for integration.

In practice, many systems will fall partly within both of the two broad categories outlined above. They may have some aspects which need to be represented by differential equations and others which involve sequential and queueing processes. Although hybrid simulation languages do exist to cater for this possibility, they are not widely available and the modeller wishing to use a special-purpose language will find himself in most cases selecting between an 'integrating' or a 'non-integrating' language.

Design Criteria for Computer Modelling

We are now at the point where the system has been analysed, we have defined the basic outline of the model and have decided on a suitable computer language to use in actually building the model. The robustness, value and flexibility of the model that is ultimately constructed depends heavily on the manner in which the model—which is presently in a symbolic form—is translated into computer code. It is now necessary to design the computer model in detail prior to actually building the model. In the following sections, we will be describing first a useful approach to laying out the structure of the model, then the concept of modular programming will be discussed. Finally, some guidelines for detailed program construction will be introduced.

TOP-DOWN MODEL-DESIGN

In Chapter 2 we discussed structure diagrams and suggested that these provide a useful medium in which to produce a visual picture of a model since they correspond well to the natural method of problem solving. That is, we start with a problem, break that up into subproblems and continue dividing each subproblem into smaller parts until we have a manageable task. This same concept (known as top-down design) can be used in designing the computer model. For example, Fig. 3.1 is a standard form of simulation-model design used by the authors. From this standard form, it is possible to derive any simulation model simply by further subdividing the four modules into small submodules. Top-down design simply involves breaking a large problem into smaller subproblems that can be dealt with

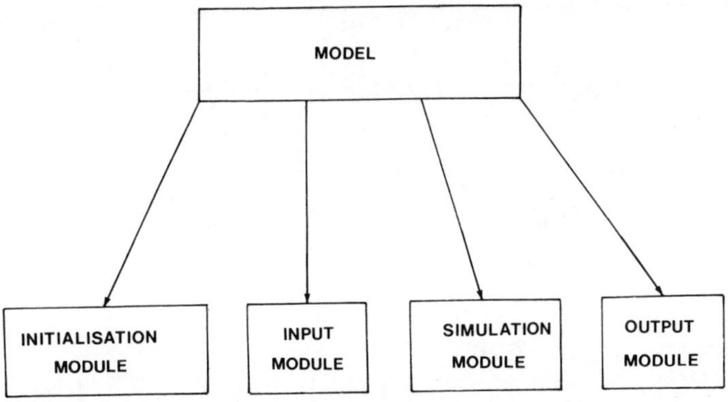

Fig. 3.1. *Standard simulation-model design.*

individually. Note that we are not involved at this stage with computer programs, but are constructing a design which can then be readily programmed. Some guidelines may be useful.

1. A systematic approach is necessary. First, specify the inputs, outputs and functions of each subproblem or module. Remember the model is to be a hierarchical structure and lateral communication should only be possible through a higher-level module.

Figure 3.2 is a structure diagram of a farm-business simulation. In Fig. 3.2, only the structure of the input module is shown in detail and this module is divided into three submodules:

(a) read in data from cards;
(b) verify data read in;
(c) print out the results of the data verification.

Each of these submodules is broken down again into yet lower-level modules. The data are read in card by card and any cards which cause a read failure due to major punching errors (such as an alphabetic character where the program specifies that a number should be found) are eliminated from

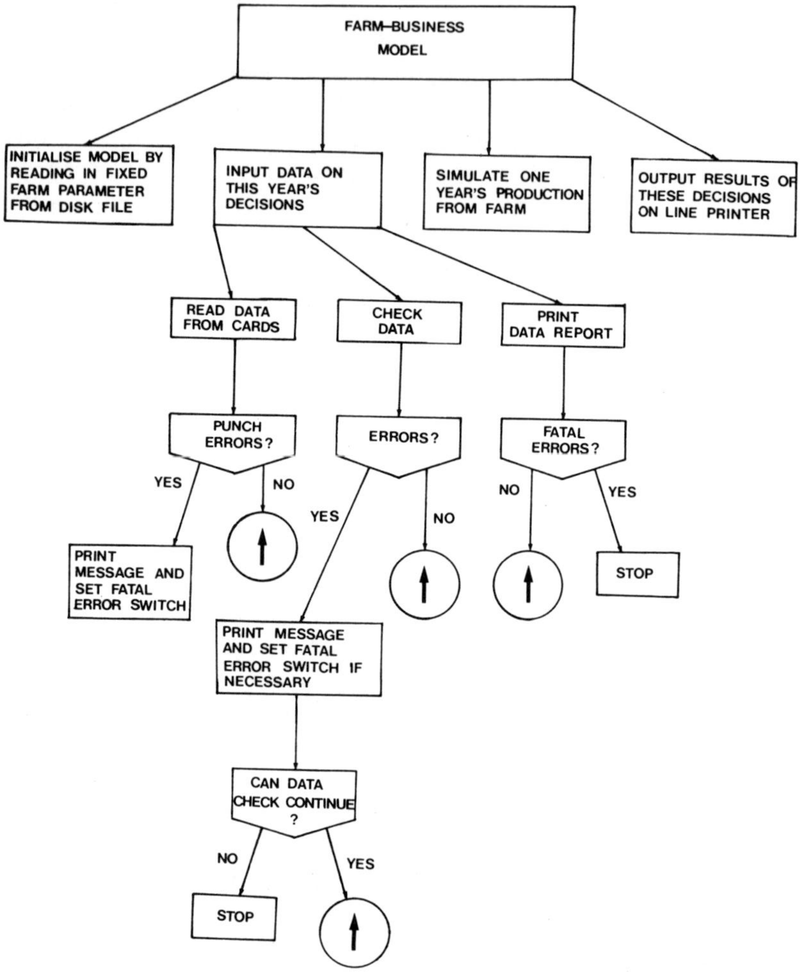

Fig. 3.2. *Partial structure diagram of a farm-business simulator—input module.*

processing and identified to the user of the model by means of a printed message. The 'fatal-error' switch—which stops execution of the model at the completion of the input module if set—is also set in this instance. Once the data are successfully read in, they are checked for consistency and correctness. Any errors detected are flagged by a printed message and, if necessary, the fatal-error switch is set. If the data error is such that further checking would be fruitless, the program stops. Otherwise the remaining data are checked and then a printout listing the data read in produced. Provided no fatal errors were found during execution of the data-input module, the model then calls the simulation module. This module, in turn, is divided into submodules (not shown in Fig. 3.2) dealing with crop and livestock activities, purchase of machinery and so on. These submodules are, in turn, served by even lower-level modules such as one to track the asset structure of the farm or a decision module which determines the loan conditions if the farm business requires outside finance. Finally, an output routine is called which can produce an option abbreviated or a detailed printout.

2. Design the model from the highest-level module down. It is often useful to produce a basic model-design and then develop in detail a design for one module or submodule (such as Fig. 3.2 where only the input module is fully defined). The modeller then programs and tests this design and, if satisfied, moves to designing in detail a further part of the model. Just as it is easier to understand a problem by slowly breaking it down into smaller pieces, so also is it simpler to build a model by building the broad outlines (or high-level modules) and filling in the details (or low-level modules) as experience and knowledge of the system develop.

Refer again to Fig. 3.2. In this case, the model-builder has decided that, as a major source of 'bugs' in models results from data input errors, he will develop and test the input module before proceeding with any of the other modules. This module can be designed, developed and tested as an independent entity. Once it is operating successfully, the remaining three modules can be built and linked to the input module. Alternatively, the model-builder could have decided to concentrate on the simulation module first. In this case, his initial design might look as in Fig. 3.3. Here the input module is reduced to its most basic function sufficient to allow development and testing of the simulation module.

Top-down design and testing has the advantage that higher-level modules (which have more use made of them) become more thoroughly tested than the lower-level modules. The model grows with the modeller's knowledge and experience, and modifications to the structure of the model, even at a late stage, are easier than if the model has been built from the lowest module upwards. Robustness and flexibility are more likely in models built using top-down concept.

MODULAR PROGRAMMING

The model is now designed as a series of high-level modules which in turn are composed of a series of submodules. This structure must now be programmed and the procedure used by the authors is to define the high-level modules as main subroutines. Within each subroutine, the next level of

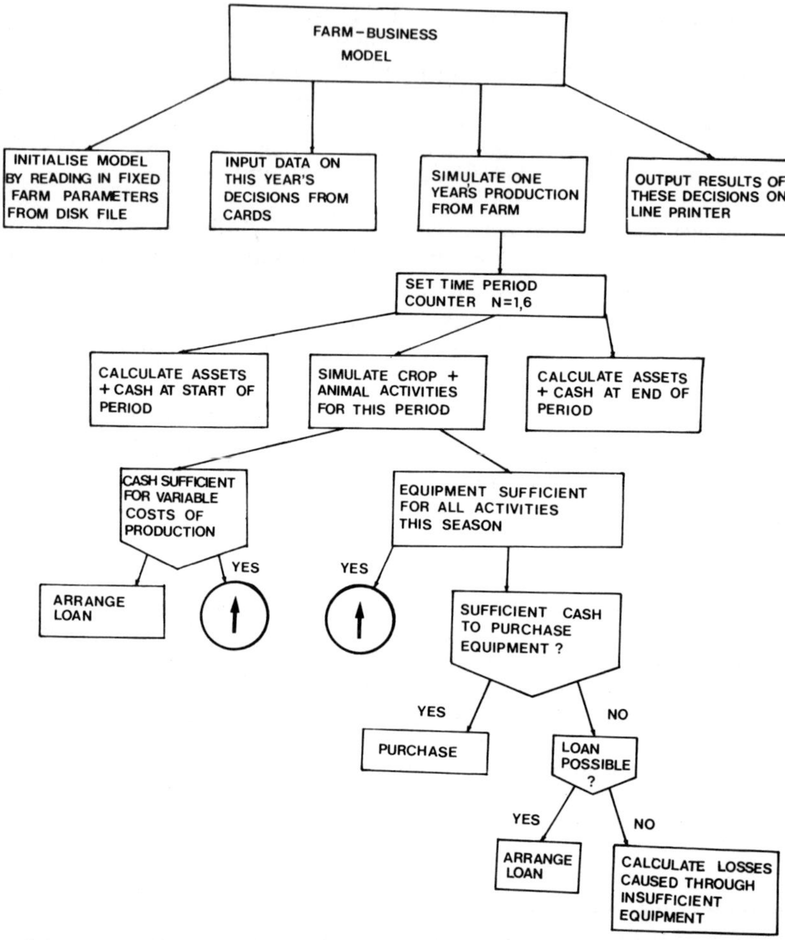

Fig. 3.3. Partial structure diagram of a farm-business simulator—simulation modules.

modules are defined as separate units within the subroutines. These submodules may be further subroutines or distinct 'blocks' within each main subroutine.

Examine Fig. 3.4. Figure 3.4a represents the mainline program of the farm-business simulator and consists simply of comment cards, a write statement to print out the version number of the program and calls to four subroutines representing the four high-level modules within the program.

(a) Mainline program

```
C   THIS SIMULATOR CONSISTS OF FOUR MAIN MODULES
C   1.  SUBROUTINE INIT — READS DATA FROM DISK
C   2.  SUBROUTINE READIN — READS DATA FROM CARDS
C   3.  SUBROUTINE SIMULA — SIMULATES ONE YEAR ON FARM
C   4.  SUBROUTINE OUTPUT — PRINTS RESULTS
C   FULL DETAILS OF EACH SUBROUTINE APPEAR IN EACH SUBROUTINE

        WRITE(6, 10)
    10  FORMAT('1VERSION 8.0 COMPILED 10/5/78')
        CALL INIT
        CALL READIN
        CALL SIMULA
        CALL OUTPUT
        STOP
        END
```

(b) Part of subroutine SIMULA

```
SUBROUTINE SIMULA                                                        ┐
  .                                                                      │
  .                                                                      │
  .                                                                      │
C   XXXXXXXXXXXXXXXXXXXXXXXXXXXXXXXXXXXXXXXXXXXXX                        │ HIGH
C   YEAR IS DIVIDED INTO 6 PERIOD                                        │ LEVEL
        DO 100 I=1,6                                                     │ MODULE
          .                                                              │
          .                                                              │
          .                                                              │
C   ADJUST ASSET STRUCTURE FOR THIS SEASON          ┐SUBMODULE           │
        DO 131 LY=1,10                              │WITHIN              │
          .                                         │MAIN                │
          .                                         │SUBROUTINE          │
          .                                         │                    │
    131 CONTINUE                                    │                    │
C   END OF ASSET STRUCTURE ADJUSTMENT MODULE        ┘                    │
                                                                         │
        CALL WORTH          ←  CALL TO SUBMODULE                         │
                               "WORTH"OUTSIDE MAIN                       │
                               SUBROUTINE                                │
                                                                         │
100     CONTINUE                                                         │
        RETURN                                                           │
        END                                                              ┘
```

(c) Part of subroutine WORTH

```
          .
          .
          .
        DO 100 MM=1,10
C   CALCULATE LONG + MEDIUM TERM LOANS OUTSTANDING
    IF (AINT(MM) .EQ.RATINT(3))ALOANM=ALOANM+ALOAN(MM)
    IF (AINT(MM) .EQ.RATINT(2))ALOANL=ALOANL+ALOAN(MM)
100 CONTINUE
C   CALCULATE TOTAL FARM CAPITAL
    TTCAP=CASH+TTBOND
    DO 101 MM=1,5
    TTCAP=TTCAP+CAPITA (MM)
101 CONTINUE
C   CALCULATE FARM EQUITY IN BUSINESS
        EQUITY=TTCAP-BLOAN-ALOANM-ALOANL-ELOAN
        RETURN
        END
```

Fig. 3.4. An abbreviated computer code for a farm-business simulator.

Figure 3.4b shows part of subroutine SIMULA which performs the annual simulation of the farm business. This main module consists of a major DO-LOOP which is used to define the simulation clock of six time periods during the year. Within this subroutine or main module, submodules may be defined. These submodules may be 'blocks' within the subroutine such as the asset-adjustment submodule in Fig. 3.4b. Alternatively, as for WORTH to calculate equity in the business they may be further subroutines (Figs. 3.4b and 3.4c).

The choice whether to make a submodule a 'block' or a further subroutine is often one of convenience. In Fig. 3.4b the asset-structure submodule is required only at the start of each season. However, the submodule WORTH is used at a number of fixed points during the simulation. It is used at the beginning and end of each year to estimate change in equity over the year, and also to evaluate whether the farm is solvent whenever a loan is required. Hence this submodule is best written as a subroutine and called whenever needed in the simulation.

The literature abounds with models that are claimed by their authors to be modular but which, on closer examination, turn out to be poorly structured and designed. The following principles will be of use in writing modular programs:

1. only one entry and one exit from each module whether high or low level;
2. limit the module in size but do not over-modularise. Low-level modules should be small, perhaps 20–40 statements in length (Shannon, 1975; Yourdon, 1975), though the exact length will vary according to circumstance. The overall size of a high-level module or subroutine will obviously depend on the number of low-level modules included.

   However, if such a module exceeds, say, 500 statements, consideration should be given to the feasibility of dividing it into two or more separate subroutines;
3. decision or truth tables can be drawn up to ascertain all possible conditions under which a module will be entered. The use of such tables can assist in ensuring that modules are not unexpectedly called during a model-run under conditions not foreseen by the model-builder.

The simulation module of the pig-budget simulation is divided into a series of lower-level modules, each dealing with a different aspect of the pig-production system. Figure 3.5 is the truth table developed for the module used to simulate the method whereby sows are introduced into the breeding

	1	2	3	4
Condition				
Wish to breed own replacements	True	True	True	True
Female replacements available	True	False	False	True
Replacements needed	True	True	False	False
Action				
Buy from outside		*		
Obtain from own stock	*			
Neither			*	*

Fig. 3.5. *Truth table for entry of sows into the breeding herd.*

herd either to replace sows culled through age or infertility or else to expand the existing breeding herd. These replacement sows may be produced from female stock bred and fattened by the farmer or may be purchased from outside the unit. Where the farmer has a stated preference for breeding his own replacements, replacements may still need to be purchased if the requirement for new sows exceeds the supply available from within the unit. Figure 3.5 shows three conditions or decision rules which must be satisfied during the execution of the replacement-sow module. The three actions which the model must take are also detailed. The table shows that only if all three conditions are satisfied can the model obtain female stock from within the unit. In all other instances, whenever female replacement stock are needed, they must be purchased from outside the unit.

The advantage of the truth table is that it forces the modeller to organise his thoughts and to lay out *all* the implications necessary for a given action in tabular format. This approach is particularly useful in complex decision problems.

PROGRAMMING CONSIDERATIONS

The final section of this chapter deals with some of the detailed points of computer programming. Elegant and careful model-designs may be ruined by careless or incompetent computer programming and the objective of this section of the text is to provide some guidelines or standards for the computer-programming stage of model-building. These standards are drawn from those used by commercial computer software developers (see Yourdon, 1975, for example) and are intended to assist in the development of reliable and useful models.

1. *Clear specifications of model-performance*

Prior to the start of the computer-programming exercise, the expected accuracy of input to, and output from, the model should be explicitly defined. This gives the model-builder and reviewers an objective base from which to assess the model. It also means that the modeller can build internal checks into the model to reduce the possibility of input data invalidating model-assumptions.

In the pig-budget simulator, key data items required are age of pigs at weaning, weight of feed fed per day, feed cost, etc. Each one of these items can be expected to fall within a known range and the model checks each data item for validity against an appropriate range. For example, although young pigs can stay with their mother for an indefinite period of time, normal management practice is to wean them, at the latest, at 10 weeks of age. The model, therefore, was built with the assumption that all pigs were weaned at 10 or less weeks of age. Hence a data item pronouncing a weaning age of greater than 10 (or less than zero!) invalidates the model assumptions and is flagged as an error. The offending value is printed out together with information on the valid range of values.

2. *Clear documentation*

One of the essentials in model-building is an accurate and lucid record of progress to date. Such a record enables the model-builder to communicate his model to potential users and is of considerable value if he has to seek assistance in the construction or use of the model. A common and useful aid to documentation is the inclusion of 'comment' cards in the body of the computer program. Such comment cards are usually bounded by minimal language syntax rules. They enable the modeller to insert verbal or diagrammatic descriptions of various parts of the model alongside the appropriate computer code. Many novice modellers ignore comment cards, intending to insert them when the model is completed. The frequent result is that late in model-development a 'bug' appears in an early piece of code. The modeller by then has forgotten the logic he used and has to waste time deciphering his original logic or else rewriting the offending code from scratch. The comments should be clear, plentiful, informative and accurate. Out-of-date misleading documentation is probably worse than no documentation at all.

Once the model is at a stage that can be used, then more formal and extensive documentation should be prepared. The following list indicates the type of documentation required if the model is to be used to any extent by other than the model-builder:

(a) program listing;
(b) variable listing and definitions;

(c) detailed structure diagrams and/or flow charts;
(d) overall description of model including concepts of model-design and theory;
(e) user's manual;
(f) programmer's manual;
(g) data sources and the manner in which data have been used in the model;
(h) contact location of the model-builder;
(i) validation results, including results reported by other users, if any;
(j) contact location of model-users plus any useful information on the outcomes of model-use.

3. *Use only standard compiler features*
The compiler, as previously explained, is a computer program itself and, like all computer programs, will have its quirks and peculiarities. It is sound modelling practice to avoid using any non-standard or undocumented features of the compiler. It is also important to check the manner in which the compiler treats operations such as mixed-mode arithmetic.

Calculations (mixed-mode calculations) involving both integer (whole number) variables and real (continuous) variables may be treated differently by various compilers. In the calculation below, the number of pigs born per day (integer variable *INUMBR*) is an integer variable calculated from the number of sows giving birth that day (integer variable *ISOW*) and the average litter size (continuous variable *ALITTR*).

$$INUMBR = ISOW \times ALITTR$$

Let the variable *ISOW* be 1 and the variable *ALITTR* be 9·6. The calculation must result in an integer result since the variable *INUMBR* is defined as integer and hence the following results are all possible:

(a) calculation is done in real arithmetic and result rounded to nearest whole number.

$$\begin{aligned} INUMBR &= 1 \times 9{\cdot}6 \\ &= 9{\cdot}6 \\ &= 10 \end{aligned}$$

(b) calculation is done in real arithmetic and result is truncated to nearest whole number.

$$\begin{aligned} INUMBR &= 1 \times 9{\cdot}6 \\ &= 9{\cdot}6 \\ &= 9 \end{aligned}$$

(c) calculation is done in integer arithmetic. Variable *ALITTR* is rounded to nearest integer value.

$$INUMBR = 1 \times 10$$
$$= 10$$

(d) calculation is done in integer arithmetic. Variable *ALITTR* is truncated to nearest integer value.

$$INUMBR = 1 \times 9$$
$$= 9$$

4. *Avoid indiscriminate program branching*

Many software developers forbid the use of 'GO TO' branches in their programs. Such statements have a proven history of causing unsuspected 'bugs' in computer programs—hence their unpopularity in the commercial world. The modular construction of computer models was described previously and it is a useful rule to ensure that no 'GO TO' statement allows the program to branch outside a given module. Branches should either be confined within a module or else go to the terminating statement of that module. This rule controls the complexity of the branching within the body of the module. In any event, beware of using 'GO TOs' to branch backwards in a program as in Fig. 3.6. Such branches can cause considerable problems in later stages of program development.

Figure 3.6 is a flow chart referring to the income-calculation submodule of the farm-business simulator described earlier in this chapter. The model-builder needed to distinguish whether income was from crops or animals but many of the calculations required in either case were the same. Instead of putting the common calculations in a subroutine, the model-builder developed the rather complex backward branching switching system illustrated in Fig. 3.6. While this system works, it is obviously clumsy and it will undoubtedly cause considerable problems at a later date when the model is further developed.

5. *Having a programming 'system'*

Have a system of naming variables and using labels. Variable names should be meaningful, particularly in languages such as ALGOL, where it is possible to define variable names of at least 63 characters. With regard to labels in FORTRAN, the authors use the scheme shown in Fig. 3.7. Each low-level module within a subroutine has a terminating statement of 1000 for module 1, 2000 for module 2, etc. Control from within module 1 can only leave that module via statement 1000. All 'GO TO' statements within that module are numbered 1001 to 1999 and all DO-LOOPS within the module end in statements 100 to 199. FORMAT statements within the

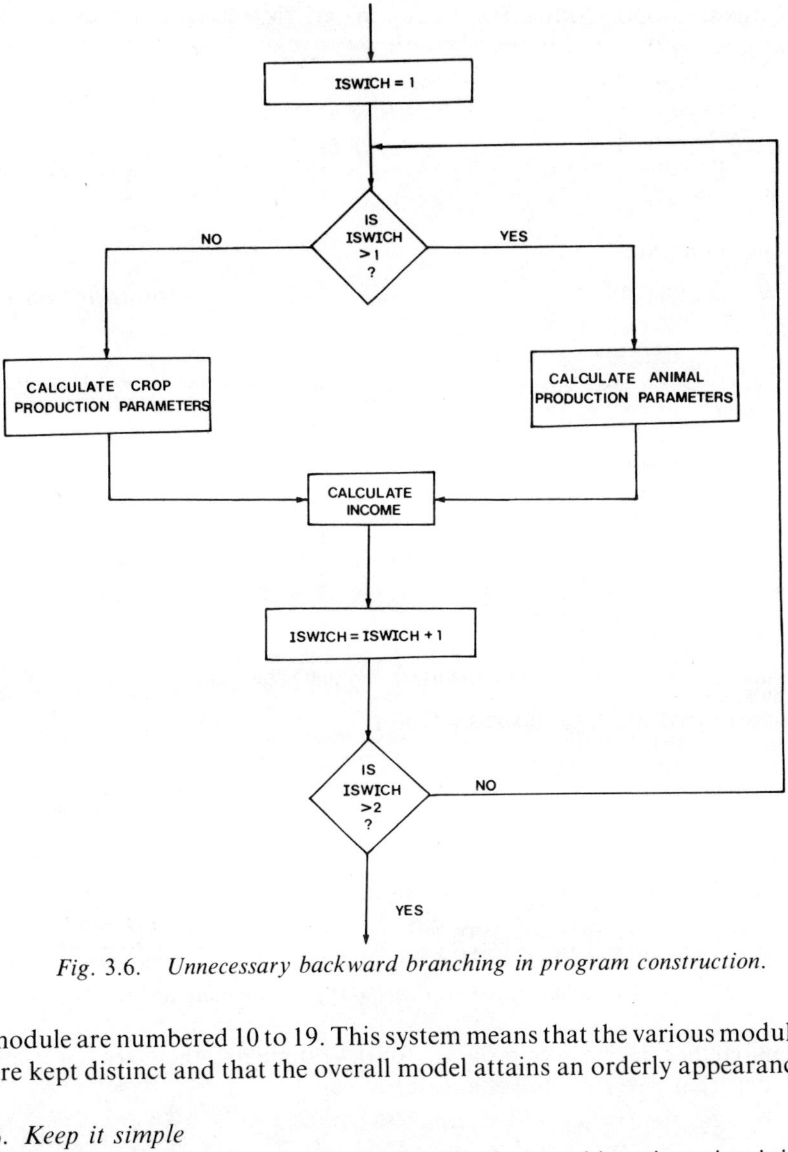

Fig. 3.6. *Unnecessary backward branching in program construction.*

module are numbered 10 to 19. This system means that the various modules are kept distinct and that the overall model attains an orderly appearance.

6. *Keep it simple*

Keep the program design simple which will reduce problems in maintaining and developing the model. This may result in some inefficiencies in the actual computer execution of the model but in general these costs will be more than compensated by reduced charges for program development and

```
C THIS SUBMODULE ESTIMATES THE # OF SPORES GERMINATING TODAY
C IF DEW IS ABSENT, SKIP THIS MODULE

      IF(DEW(IDAY) .LE.O.O)GØ TØ 1000                    GO TO labels in
          .
          .
          .
      IF(GR)1001,1000,1002                               this module 1000–1999.
          .
          .
          .
 1001 CONTINUE
      DO 100 J=1,4                                       DO-LOOP labels in
          .
          .
          .
      DO 101 K=1,7                                       this module 100–199.
          .
          .
          .
  101 CONTINUE
          .
          .
          .
  102 CONTINUE
          .
          .
          .
 1000 CONTINUE
C THIS SUBMODULE ESTIMATES SPORE PRODUCTION WHICH
C IS DEPENDENT ON AVERAGE AMBIENT TEMPERATURE THIS
C WEEK
C CALCULATE AVERAGE TEMPERATURE
          AVTEMP=0.0
          DO 200 J=1,7
          .                                              DO-LOOP labels in
          .                                              this module 200–299.
          .
      200 CONTINUE
          .
          .
          .                                              GO TO labels
          IF(AVTEMP.GT.25.)GØ TØ 2001                    in this module
          IF(AVTEMP.LT.10.)GØ TØ 2002                    2000–2999.
```

Fig. 3.7. A labelling system for modular programming in FORTRAN.

maintenance. Design the program for development—there are few useful models that never require changes.

Summary

The writing of the computer model can be a relatively straightforward process. The language selected should be well supported by the computer centre and should suit the type of model being developed. The actual writing process should follow a system—both in the syntax and the structure of the program. The program should be divided into clearly distinct modules which have one entry and one exit point. The top-down design of the program which follows the structure-diagram concept outlined in Chapter 2 has much to recommend it. With the alternative bottom-up design, testing cannot be satisfactorily accomplished until completion of the entire model. A strict modular approach to the programming of the model will produce the kind of model which can be adapted and expanded as needed.

Workshop

1. Describe a continuous and a discrete system with which you are familiar and set out partial structure diagrams with particular reference to the modular structure of a simulation subroutine representing each system.
2. Establish a data-checking procedure for the above systems model which will indicate errors when data-input items occur outside specified ranges.

References

Charlton, P. J. (1971). Computer languages for system simulation. *In:* Dent, J. B. and Anderson, J. R. (eds), *Systems Analysis in Agricultural Management.* Sydney: Wiley.

Shannon, R. E. (1975). *Systems Simulation: The Art and the Science*, Englewood Cliffs: Prentice-Hall Inc.

Teichroew, D. & Lubin, J. F. (1966). Computer simulation—discussion of the technique and comparison of languages. *Commun. A.C.M.*, **9,** 723–41.

Yourdon, E. (1975). *Techniques of Program Structure and Design.* Englewood Cliffs: Prentice-Hall Inc.

Further Reading

Chapas, L. C. (1970). Computer languages for model building. *In:* Jones, J. G. W. (ed), *The Use of Models in Agricultural and Biological Research.* Hurley: Grassland Research Institute.

Claney, J. J. & Fineberg, M. S. (1965). Digital simulation languages: A critique and guide. *Proc. A.F.I.P.S., Fall Joint Computer Conf.*

Dimsdale, B. & Markowitz, H. M. (1964). A description of the SIMSCRIPT language. *I.B.M. Systems J.*, **3,** 57–67.
Llewellen, R. W. (1965). *FORDYN—An Industrial Simulator*. Raleigh, N. C.
Pugh, A. L. (1963). *Dynamo Users' Manual.* Cambridge, Mass.: MIT Press.
Radford, P. J. (1970). Some considerations governing the choice of a suitable simulation language. *In:* Jones, J. G. W. (ed), *The Use of Models in Agricultural and Biological Research.* Hurley: Grassland Research Institute.

4

Stochastic Specification

The Case For and Against Stochasticity in Modelling

So far we have talked in various sections about stochastic events and their use in simulation models. The need to specify stochastic variables arises partly because of the conceptual boundary we have drawn round the system to be modelled. It is often the case that, in viewing a system at a number of different levels of organisation, what may appear to be unexplained or even random variation at one level is explicable at another. The term 'stochastic' can then be used to describe both unexplained events (as far as this model is concerned) and events which are truly random.

Many texts on simulation (see Mihram, 1972 for example) emphasise the need for the inclusion of stochastic elements in simulation models. Mihram (1972), in fact, claims that it is virtually impossible to build a realistic simulation model unless stochastic elements are in the model. Predictive models, in particular, are concerned with the future, and the systems of interest to agricultural scientists almost invariably involve an uncertain and risky future. This is especially so in management-orientated applications, where managers have to take decisions, the outcome of which will not be known until some future date. Anderson (1976) pointed out that modellers have a responsibility to provide decision makers with guidance and data which recognise the full extent of risk inherent in any policy. Therefore, he suggested that where a model was being used for, say, planning buffer stocks of food supplies, planners could not afford to work with only single valued estimates of intrinsically risky phenomena.

Where models are being used in a decision-support role, uncertainty must be explicitly included in the model-structure so that the model, in its totality, may reflect the degree of understanding of the real system. Deterministic models can only yield information on the mean response of the real system to a change in one of its variables. Good decision making will require more information than simply a knowledge of the average or most likely response.

On the other hand, it is important that the stochastic features of the model are incorporated in a form that is acceptable to, and understandable by, model-users. Where a predictive model is used to generate a farm or enterprise budget, many users (such as farmers or farm consultants) will prefer to use such a model deterministically. They will investigate the stochastic elements of the budget by a series of runs with different price and yield levels and subjectively evaluate the results. The inclusion of stochastic elements directly in such models often creates confusion and reduces the acceptability of the model to the user. In an explanatory simulation study of a biological system, stochastic elements in a model may contribute little to the understanding of the real system.

We have delayed detailed consideration of stochasticity until this stage for two reasons. Firstly, we believe the principles of model-building can be better taught using deterministic models initially and then introducing stochastic elements where deterministic models appear inadequate. Secondly, we disagree with the premise that stochastic elements are always essential in advanced simulation work. Stochastic elements are important in bioeconomic simulation modelling but they should be introduced with caution. Useful models can be built on a deterministic basis and the modeller should beware of using doubtful probability data or of making his initial model too complex simply to accommodate uncertainty. Again it is important to consider the objectives of the modelling exercise. It is better to have a simple deterministic model which meets the stated objectives and which will be used, than to have a complex stochastic model which is too expensive, too clumsy or too obscure to be satisfactory to the user.

There are two sources of uncertainty which need to be considered in agricultural simulation modelling:

1. The data used to develop relationships for the model will never be error-free. Each relationship, therefore, will always have an error term representing, *inter alia*, the effect of variables not explicitly considered, the difficulty of precise measurement of biological and economic data, lack of fit of the statistical function and inherent biological variation. The modeller may, in certain instances, make the positive subjective decision that the relationship without an error term is adequate for his purposes. It does appear, however, that frequently the error term is omitted without adequate consideration of its relevance.

2. The environment in which the real system operates is usually unpredictable and uncontrollable; it is, therefore, a major source of uncertainty.

The uncertainty associated with the exogenous variables of a model is compelling and is the major issue of this chapter.

Simulation models explicitly include time and, as we have seen, one of the most important aspects of a simulation 'run' is the information the model provides on the changes occurring in status and output variables over time. These changes are partly determined by the values of the exogenous variables in each time period so it is obvious that realistic values for the exogenous variables must be provided to the model for each time period. By 'realistic' values, we do not mean that we are attempting to predict exact values for future events. Rather we intend that the time series of exogenous variables used in the model should be representative of the environment taking particular account of known patterns in, and interactions amongst, variables. Such a representative series may be obtained in either of two ways:

1. by using historically recorded time-series data for an exogenous variable in the anticipation that, for this variable, the past is a reasonable indication of what might be expected in the future. This is not to suggest, for example, that the rainfall sequence for any particular locality will ever occur again, but that it is indicative of what might be expected;
2. by providing information structures in the model which are capable of generating representative time-series data. These data of course may never occur in the real world exactly in the sequence generated by the model.

Both procedures can provide acceptable environmental data for simulation models. Each approach has its advantages in different circumstances and the model-builder should be aware of the issues involved in deciding which to use in his model.

It is worth noting at this point that the output from a stochastic model can be compared with that of a real system using conventional statistical methods since both sets of output are random variables. Much use can be made of this both in validation of, and in experimentation with, computer models. (These topics will be discussed in detail in Chapters 5 and 6 respectively.) If a model is deterministic, there will be no true 'experimental error'—and hence no error (or no appropriate measure of error) for computation of *F* tests in analyses of variance or confidence limits in regression.† It is obviously possible to highlight mean differences between two sets of output using a deterministic model but note Anderson's (1976) cautionary words on the use of averages for decision making in an uncertain world.

THE USE OF HISTORICAL TIME-SERIES DATA

In many cases the direct use of historical data has attractions. For example, the use of rainfall records as data for a simulation model is frequently convenient, as such records are often readily available and may be used without the trouble of major statistical analysis. In some cases, however, historical data may not be available; and even where records are available, they are often of limited length compared to the periods of time which we may wish to simulate. Further, the use of historical records prevents us from exploring the effects of events which have not been recorded previously or the effects of different sequences of events.

Some of these limitations may be overcome fairly simply. Rearrangement of the data will provide different sequences of values, but this incurs the risk that the rearranged data may fail to display any autocorrelation which existed in the original data.‡ Where a long series of records is available, choice of different starting dates for simulation experiments covering shorter periods may be an alternative. Obviously, neither of these approaches is suitable where only short historical time series are available nor do they overcome the problem of testing the simulation model under conditions which have not been observed during the recorded period. Phillips (1971) provides a useful summary on the limitations of the use of historical data from a statistical standpoint. We should note one further problem associated with the use of historical data and that is the physical problem of computer storage and retrieval of large amounts of historical data. If a model requires a time-series observation for each week of a 20-year 'run', a data file of over 1000 records must be created

† There may be an apparent 'error': in an analysis of variance this would represent lack of fit of the linear statistical model.

‡ Autocorrelation of observations in a time series means that successive observations in the series are dependent to some extent. A simple rearrangement of an autocorrelated series will result in the loss of the relationship between successive observations and, therefore, the rearranged series will not be a valid alternative sequence to the original series. The issue of autocorrelation will be discussed later in this chapter.

for, and accessed by, the model. Where several exogenous variables are involved, the use of a large data file in association with the model may not be acceptable or efficient.

In spite of these disadvantages, historical data are useful particularly in the early stages of model-development. The model can be developed and tested using a limited historical time series and later modifications can involve the development of appropriate stochastic time-series generators (see, for example, Ritchie, Dent and Blackie, 1978).

GENERATION OF TIME SERIES IN THE MODEL

The alternative to the direct use of historical records is to generate values of the exogenous variables as stochastic time series drawn as samples from probability distributions. Remember that a value for each exogenous variable must be provided for each time interval in a manner which is representative of the variable in the real system. The generated time series then must be an acceptable sample from the overall population for that variable. In constructing simulation models of agricultural systems we are frequently beset with seasonal or annual cycles in the environment and these must be represented in any time series that is generated by the model.

One acceptable procedure to follow in generating time-series data is as follows:

1. Examine the available historical records for the exogenous variable under study.
2. Collate all the data for each time interval as a probability distribution. For example, if we are concerned with rainfall and the selected model time interval was a calendar month, we would first assemble the historical records for rainfall in January, February and so on, into ordered arrays. The rainfall data for each month are then described in the form of a probability distribution.
3. A random selection from the January probability distribution would be the first element in the generated time sequence while the second element would be a random value from the February distribution. We can proceed in this manner calling a random value in turn from each of the twelve monthly distributions to give a monthly rainfall pattern for one year and a value for total annual rainfall. Obviously, the annual total must be within accepted limits or the generated pattern would not be satisfactory. If a simulation over a longer period than 1 year is needed, the above process is repeated with new random selections being made in order from the

appropriate distributions. This procedure permits the generation of a time series of any required length. The seasonal trends in the recorded data will be reflected in the generated series by virtue of the different monthly probability distributions.

Autocorrelation

The procedure outlined above may not be satisfactory in all cases. The implicit assumption in this method is that individual time-series values (monthly rainfall in the example above) are independent. Many biological and economic time series consist of individual observations which are not independent but are autocorrelated.

Where autocorrelation is present then successive observations in the time series are linked in some manner. With positive autocorrelation, the second (or some later observation) tends to resemble or repeat the first observation. To illustrate this in the rainfall context, autocorrelation may be strongly present between months so that in a year when low rainfall is experienced in January there is a strong chance that less than average rain may fall in February. If this autocorrelation is not reflected in the generated series, the real environment is not being reproduced faithfully. Where autocorrelation is displayed between elements in the recorded data then the simple sequential independent random selection of values as illustrated above may be unacceptable.

TESTING FOR AUTOCORRELATION

Autocorrelation tests (sometimes called serial correlation tests) are an important class of statistical tests used to check the assumption of independence between individual observations in data records. A variety of tests for autocorrelation have been developed with important differences in properties between them (Johnston, 1972). One of the most important characteristics is the circularity of the test. Where data exhibit a trend, the values of the first and last observations may dominate the circular tests so as to render autocorrelation between the remaining observations non-detectable. A circular test should not be used with data that exhibit a trend. Trends of an annual or seasonal type may be removed by deflating the series or by fitting a curve to the trend and taking the differences between the observed values and the curve as being the adjusted series of data (Phillips, 1971). Autocorrelation tests applied to the adjusted series may reveal forms of interdependence other than those associated with the trend or cycle.

One of the simpler non-circular autocorrelation tests to use is the 'runs test'. Data are presented as a series of positive or negative deviations from a measure of central tendency (e.g. the mean or the median). Any zero deviations are deleted from the sequence and the number of positive (n_1) and negative (n_2) deviations are calculated ($N = n_1 + n_2$). A 'run' is a series of one or more positive or negative deviations. The number of 'runs' of values of like sign are then compared with that which might be expected in a random (non-autocorrelated) series. The observed number of runs (u) is compared with the expected value ($E(u)$) by the statistic Z where

$$Z = (u - E(u))/\sigma_u$$

$$E(u) = 1 + (2n_1n_2)/(n_1 + n_2)$$

and

$$\sigma_u^2 = \frac{(2n_1n_2 - n_1 - n_2)(2n_1n_2)}{(n_1 + n_2)^2(n_1 + n_2 - 1)}$$

Where n_1 and n_2 both exceed 10, the statistic Z is distributed as a standard normal deviate and may be tested by comparison with a table of areas under the normal curve. For a two-tail test, the null hypothesis is rejected and hence autocorrelation assumed with 95 per cent confidence if

$$|Z| > 1{\cdot}96 \text{ for } p < 0{\cdot}05$$

For smaller samples, reference to special tables is required (see Conover, 1971).

Table 4.1 illustrates the results from the comparison of a series of monthly rainfall data for four sites in New South Wales, Australia. Both the mean and the median are tested and both indicate evidence for significant autocorrelation amongst months.

ACCOUNTING FOR AUTOCORRELATION

When autocorrelation is detected in a time series, the implication for the modeller is that the straightforward procedure of determining a probability distribution for each time interval (say, each month), and drawing random variates sequentially from these to produce the required generated series, must be foresaken. Autocorrelation means that there is, in the base data, strong between-period interrelationships such that

$$X_t = \mathrm{f}(X_{t-1}, X_{t-2} \ldots X_{t-n})$$

where X_t is the value of the variate in period t, X_{t-1} is the value of the variate

Table 4.1. Runs tests for a series of monthly rainfall observations (after Filan, 1975.)

	Deviations from means Station				*Deviations from medians* Station			
	1	2	3	4	1	2	3	4
Number of observations	911	911	911	635	911	911	911	635
$E(u)$	413	429	405	289	456	456	455	311
σ_u	13·61	16·16	13·37	11·40	15·07	15·07	15·06	12·44
u	391	405	377	249	415	420	437	273
Z	−1·55	1·69*	2·09**	−3·49***	−2·72**	−2·39**	−1·20	−3·06***

* $p \leq 0{\cdot}10$
** $p \leq 0{\cdot}05$
*** $p \leq 0{\cdot}01$

in the preceding period and X_{t-n} is the value of the variate n time periods ago.

It will normally be possible to establish the form and the parameters of the above relationship by assembling the base data into time periods and carrying out regression analysis. The simplest format would be a linear, one period-lagged regression of the form

$$X_t = a + bX_{t-1} + e_t$$

Phillips (1971) argues that the parameters for this relationship are best determined by adjusting the base data to differences from the mean in each period. The format then becomes

$$(X_t - \bar{X}_t) = a + b(X_{t-1} - \bar{X}_{t-1}) + e_t$$

The error term should be normally distributed with a mean of zero and constant variance. Error is then estimated by a stochastic sample from the appropriate normal distribution.

The total procedure would then be to determine the first value in the generated stochastic series (X_1) from its probability distribution and then to use the established recursive regression together with a random error to calculate the next value in the series (X_2). Following this, the relationship between X_2 and X_3 will be used in a similar manner to determine the value of X_3, and so on.

In the situation where the variate is monthly sunshine hours and the simulation begins in January, the steps would be as follows:

1. From the probability distribution for January, randomly select a value for the number of sunshine hours.
2. Using the relationship between February and January sunshine hours (assuming a linear, one period-lag) and the value for January determined in 1. above, calculate the value of February sunshine hours.
3. Proceed in a similar fashion through to December.
4. Using the value for December sunshine hours and the recursive regression between January and December, calculate the value of January sunshine hours in the second year of the simulation. Note that this will be different from the first year figure because the first year December figure is taken into account and because the error term in the relationship is randomly drawn.
5. Proceed for each month over the full number of years of the simulation.

Incorporating linear one period-lag recursive regression into the model-structure is a major extension of modelling complexity and often the modeller will seek ways to avoid it (see later in this chapter). Further complexity of a daunting nature is involved in a situation where more than a one period-lag needs to be taken into account or where a linear function cannot be assumed.

As noted by Phillips (1971), the model-builder has other options than to use the recursive regression approach to incorporate autocorrelation: Markov processes, for example, may be applied to achieve similar results. No lesser degree of complexity may be expected from these other approaches.

Generating Random Variates

Whether or not autocorrelation is present, stochastic elements in simulation models are created through the selection of a random variable from some defined probability distribution.

There are, therefore, two steps in the creation of any stochastic event:

1. obtaining a random number;
2. using this number to select a particular value from a defined probability distribution.

In the remainder of this chapter we will discuss first the basic concepts of probability, then the use of a pseudorandom-number generator to obtain random numbers and finally how to use random numbers to create a stochastic event.

PROBABILITY FUNCTIONS

Statistical theory is built on a set of axioms or laws of probability. For example, the sum of probabilities of all possible but mutually exclusive events must be 1·0. These laws of probability are outlined in a variety of statistical textbooks. A number of algebraic forms of equation have been found to have the mathematical properties required by the laws of probability, given that appropriate values are selected for equation parameters. Each such algebraic form represents a 'family' of probability distributions or probability density functions: a specific distribution is defined when values are assigned to, or estimated for, each parameter of the distribution. Particular families of distribution (that is, forms of equation) which are commonly applied in statistical and/or simulation work include those known as the normal (or Gaussian), uniform, gamma, beta, Weibull and exponential distributions. The text by Johnson and Kotz (1970) presents a variety of families of distribution, a detailed discussion of statistical properties, potential uses and methods for estimating parameters.

If the data on the stochastic elements of a model appear to conform to one of the known families of frequency distributions, the parameters and algebraic components of the appropriate function can be built into the model. Selection of a suitable family of distributions can be guided by the following considerations:

1. The nature of the process which the stochastic values reflect. Many of the common families of distributions have been developed in order to represent particular processes. Johnson and Kotz (1970) discuss the theoretical or physical nature of processes which are likely to generate data corresponding to different families of distribution.
2. The shape of the graphed frequency distribution can be compared to that of a known family of distributions. Again, texts such as that of Johnson and Kotz (1970) or Shannon (1975) are helpful, as they illustrate the range of forms which various distributions may take when graphed.

   Very often, if an appropriate distribution is not defined by the theory of the process which generates the data, two or more forms of distribution may appear to have a potential for providing an adequate description of the data. In such cases it may be desirable to estimate parameters for each form of distribution, and apply a statistical test for agreement between the data and the estimated distribution. In such cases, the Kolmogorov–Smirnov test may be appropriate if a large sample of data is available, while for smaller samples the chi-square test may be useful. These tests are described in many elementary and intermediate statistics textbooks (see Conover, 1971, for example).
3. Computational convenience may be a deciding factor where alternative forms of distribution describe the sample data distribution equally well: some families of distributions may be used conveniently and economically in simulation, while others are rather difficult and/or costly to use as stochastic-variable generators. The uniform, normal and Weibull distributions are convenient, but others, such as the gamma distribution, may present considerable difficulties in computation.

Some authorities find that the use of specified families of distribution is undesirable. This may be because

1. the data will not fit any of the common families satisfactorily;
2. problems of selecting random values occur.

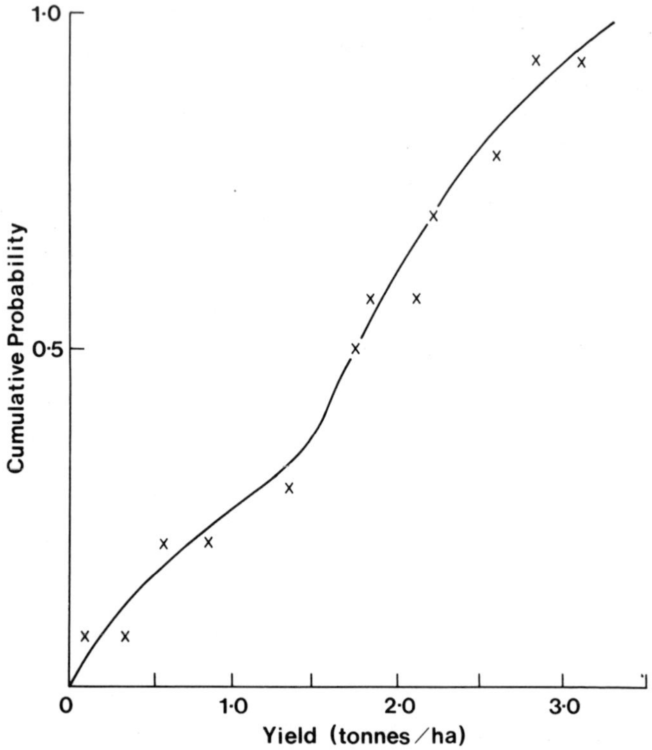

Fig. 4.1. *Cumulative probability distribution for wheat yields.*

For any but variates from normal, Weibull or exponential distributions, the latter difficulty can prove decisive. Recourse may then be made to the cumulative distribution function (CDF) (Phillips, 1971) which, in our opinion, provides the simplest and most effective method for generating random variables which can be used in model-building.

Table 4.2 presents data on average wheat yields to be used in the farm-business simulator mentioned in Chapter 3. Rather than attempt to fit the limited data available to one of the families of distributions, a cumulative distribution function was drawn as shown in Fig. 4.1. This figure was in fact drawn free hand and the data presented to the model as pairs of values read from the axes. Random selection of yield proceeded by the determination of a uniform random number between 0 and 1 and the subsequent matching of this against the appropriate value on the horizontal axis. Appropriate linear interpolation between successive pairs of values is, of course, a necessary but minor programming chore.

Table 4.2. Average wheat yield for a case property

Year	*Wheat yield (tonnes/ha)*	*Year*	*Wheat yield (tonnes/ha)*
1963	2·86	1970	1·57
1964	2·86	1971	1·64
1965	0·10	1972	0·55
1966	3·38	1973	2·46
1967	0·51	1974	2·46
1968	1·75	1975	1·56
1969	1·13	1976	2·46

The approach is quite general and can be applied to any data. For those who prefer a more precise determination of the cumulative distribution function, smoothing techniques are available based on the concept of grafting a series of polynomial functions together (Phillips, 1971). Alternatively it will often be possible to fit a simple function form to the cumulative distribution data (e.g. a simple polynomial or logarithmic function). This has the advantage of avoiding the need to present paired data from the cumulative distribution function in the model: in this case the function is built into the model and any uniform random variates applied to function will provide values of the exogenous variable.

THE PSEUDORANDOM-NUMBER GENERATORS

Once the probability distribution has been defined, it is then necessary to select random variables from this distribution. In the case of a CDF or a uniform distribution, the pseudorandom-number generator available on most computers may be used directly to generate random variates. For other distributions, a selected uniform random number (or numbers) is transformed in some manner in order to generate a random variate from the appropriate distribution. More will be said about this shortly, but first we will consider how to obtain a uniform random variate for use in a simulation model.

Large tables of random numbers are published and subsets of these tables are to be found in most statistics textbooks. However, where large samples of random values are likely to be required, storage of such tables in computer memory can be expensive and cumbersome. The alternative, which is commonly adopted for computer-based simulation work, is the uniform pseudorandom-number generator. Nearly all modern computers have these available in a form which makes their use extremely easy.

For example, in Burroughs B6700 FORTRAN programmers may obtain a uniformly distributed random value X by a statement

$$X = \text{RANDOM(NSEED)}$$

RANDOM is an intrinsic function supplied as part of the computer software. The variable 'NSEED' is, as we shall see, ultimately supplied by the program user and is used to generate a value X which is a random variable selected from the uniform distribution. The values of X are generated from a range between 0 and 1.

A pseudorandom-number generator uses a formula which will produce a sequence of numbers from a uniform probability distribution which *appear* to be random in the sense that no pattern is detectable in the sequence without knowledge of the underlying formula. However, those who know the formula are able to reproduce the same sequence at will; or to produce a different sequence whenever this may be required. The sample of numbers generated (the values of X in the above example) depends on a 'seed' or starting value (the variable '*NSEED*' in the above example). Usually this needs to be supplied by the user/programmer, although in many cases if this is not done the computer supplies a default value. Some care in the choice of a seed is essential as 'bad' seeds may lead to samples which are not random.†

Shannon (1975) describes in some detail the various methods which can be used to generate pseudorandom numbers and provides listing of suitable FORTRAN functions for this purpose. For most modellers reliance on the pseudorandom generators available as part of local computer software will be adequate. However, some research into the appropriate choice of seed for any software-supplied pseudorandom-number generator is to be recommended.

CREATING A STOCHASTIC EVENT

Once we have defined the appropriate probability distribution and selected a pseudorandom number, the remaining problem is how to use this random variate to simulate a stochastic event in the model. The simplest case is that of the uniform distribution.

Consider the situation in the barley-leaf-rust model where we wish to simulate the effect of dew on spore germination. Only if dew is present do spores germinate, and at this stage of model-development we assume that the probability of dew being present on any day is 0·5 and the probability density

† Many FORTRAN manuals unfortunately fail to indicate the appropriate criterion for a 'good' seed. This criterion depends on the particular method being used and frequently, on the word-length of the compiler involved.

function is uniform. The FORTRAN code necessary to model the effect of dew on germination would be

```
C VARIABLE 'NSEED' IS SUPPLIED TO MODEL AS
C SEED FOR FUNCTION 'RANDOM'. FUNCTION RETURNS
C VARIABLE 'RN' AS A RANDOM VARIABLE (0–1)
C PROBABILITY OF DEW = 0·5
C IF DEW ABSENT, # OF SPORES GERMINATING TODAY
C (VARIABLE 'NSPORE') IS ZERO
      RN = RANDOM (NSEED)
      NSPORE = 0
      IF (RN. GE. 0·5) NSPORE = GERM * SURVIV
```

A similar procedure can be used in the case of a CDF. Refer back to Fig. 4.1 and you will notice that the vertical axis has a range from 0 to 1—the same range as the uniform pseudorandom numbers. It is therefore, only necessary to store the values on the horizontal axis (in this case yields) in an array and select from this array according to the random number generated.

Consider the case in Fig. 4.1 where we wish to establish the yields which occur with a probability 0·1, 0·2, 0·3 . . . 1·0. We would have to create an array with 10 storage positions and, in the case of Fig. 4.1, the values stored in this array would be as follows:

Probability	Array *Position*	*Value*	*Probability*	Array *Position*	*Value*
0·1	1	0·28	0·6	6	1·83
0·2	2	0·61	0·7	7	2·20
0·3	3	1·20	0·8	8	2·34
0·4	4	1·52	0·9	9	2·62
0·5	5	1·75	1·0	10	3·18

(The reader should draw horizontal lines from the vertical axis at points 0·1, 0·2, and so on, to meet the CDF. These lines are then projected at right angles to meet the horizontal axis to give the values above.)

A random number of 0·1 would then give a yield of 0·28 tonnes/h, and a value of 0·6 would represent a yield of 1·83 tonnes/h. Obviously most of the random numbers selected will represent probabilities intermediate between those in the table above. In this case, all that is needed is to interpolate linearly between the two closest values. For example, a random number selected of 0·843 would involve interpolating between the yield 2·34 (probability 0·8) and the yield 2·62 (probability 0·9), as:

$$\text{Yield} = 2{\cdot}34 + 0{\cdot}43(2{\cdot}62 - 2{\cdot}34) = 2{\cdot}46$$

(Note that the use of a function fitted to the CDF data avoids the need for interpolation).

If the distribution to be used is not in CDF form or else is not uniform, then the model-builder will need to include a subroutine or procedure to convert the uniform random numbers selected into the appropriate probability density function. It is not necessary for the modeller to develop his own subroutines as suitable ones may be found in the text by Shannon (1975) or Naylor, Balintfy, Burdick and Chu (1966). However, a word of caution is necessary—test any such subroutine thoroughly before incorporating it into the model. Occasionally these routines do not work as specified because of one or more of the following:

1. intrinsic errors in the initial subroutine;
2. typographical errors in the documented subroutine;
3. incompatibility between computers.

Some Final Words

The understanding of the use of stochastic variables is a most important part of model-building. This chapter has discussed the arguments both for and against stochastic models and has illustrated a variety of ways in which random variables may be generated in a model. Of particular importance in modelling biological and economic systems is the problem of autocorrelation and the effect this may have on the time interval selected for the model.

In the presence of autocorrelation, and when we do not wish to use the recursive mechanisms outlined in this chapter, there are still some possible alternatives which, however, involve some compromise and which may influence the accuracy of modelling. Autocorrelation in historical time series is related to the time interval of the series. For example, in a time series of daily rainfall, strong autocorrelation might be expected (as one rainy day is likely to follow a previous rainy day). If the time interval was on an hourly basis, even stronger autocorrelation among successive readings would be expected. It follows that a lengthening of the time interval will weaken the autocorrelation: there is likely to be less relation between monthly rainfall than hourly rainfall. Where the selected d*t* for the model (see Chapter 2) involves the use of stochastic generators based on a time series in which autocorrelation is present, one possible approach would be to extend the length of the d*t* for the model. The time-series data can then be amalgamated to correspond with the new d*t* with the likelihood of diminishing or eliminating autocorrelation. As a general rule, therefore, the final decision on the size of d*t* should not be made prior to a consideration of the time series available for the stochastic generators and before appropriate autocorrelation tests have been carried out. It may well be that the model-builder considers that an extension of the basic time interval, presumably associated with some loss in model-precision, is worthwhile to avoid the complexities of accommodating autocorrelation or the inaccuracies of ignoring it. There will be no obvious right answer and the choice will depend on the model-builder's judgement. Once again he will have to try an alternative, examine the results and if necessary modify or improve the existing model-format.

Workshop

1. Take a time series of monthly rainfall records and test for the presence of autocorrelation using the runs test.
2. Create cumulative probability functions for each month of rainfall and program a model-section which will produce random monthly rainfall values over a period of 10 years. Select model-time steps (dt) which first avoids incorporation of recursive procedures and then includes them.

References

Anderson, J. R. (1976). Essential probabilities in modelling agricultural systems. *Agr. Systems*, **1**, 219–32.

Conover, W. J. (1971). *Practical Non-Parametric Statistics*. New York: Wiley.

Filan, S. J. (1975). Unpublished M.Sc. Thesis, University of New South Wales, Australia.

Johnson, N. L. & Kotz, S. (1970). *Continuous Univariate Distributions*. Boston: Houghton Mifflin Co.

Johnston, J. (1972). *Econometric Methods*. New York: McGraw-Hill.

Mihram, G. A. (1972). *Simulation: Statistical Foundations and Methodology*. New York: Academic Press.

Naylor, T. H., Balintfy, J. L., Burdick, D. S. & Chu, K. (1966). *Computer Simulation Techniques*. New York: Wiley.

Phillips, J. B. (1971). Statistical methods in systems analysis. *In:* Dent, J. B. & Anderson, J. R. (eds.), *Systems Analysis in Agricultural Management*. Sydney: Wiley.

Ritchie, I. J., Dent, J. B. & Blackie, M. J. (1978). Irrigation management: an information system approach. *Agr. Systems*, **3**, 67–74.

Shannon, R. E. (1975). *Systems Simulation: The Art and the Science*. Englewood Cliffs: Prentice-Hall Inc.

5

Model-Evaluation

Once the model is at a stage where it can generate apparently useful output we need to move to the verification and validation stages of model-development. These processes, it will be found, continue throughout the rest of the model-building and usually through to model-application. In practice the two procedures are quite distinct and have different purposes; it is essential that we make the distinction clear.

The problem of verification, at its fundamental level, involves philosophical questions regarding the concept of truth and the manner in which we can establish perfect representation (i.e. what, in truth, is the real situation and how can this be modelled?). In practical terms this can only be achieved by selecting criteria by which the model can be compared with current understanding of reality. Suitable criteria are not always obvious or easily quantifiable and the practical issues of verification are, therefore, as elusive as the philosophical arguments on the nature of truth (Naylor & Finger, 1967).

Fortunately, in practice, we are rarely concerned with absolute truth nor in perfect representation with our models. If we accept that a functional model is an abstraction of reality and an approximation to the functioning of the real system, then the problem is not to establish the 'truth' of the model in any absolute sense but to determine whether the model we have constructed is an adequate representation for our purposes. If we build a model of an irrigation system which includes plant-growth and soil-moisture balance modules, this model may be quite adequate for assessing alternative irrigation-policy decisions. On the other hand, this same model could be of limited value in predicting the detailed physiological responses of individual plant species to variations in soil-moisture balance.

The adequacy of a model then must be seen in relation to its purpose and not from an absolute viewpoint. We are then mainly concerned with a process of validation rather than verification. For this purpose, we can avoid the philosophical discussions and attempt to locate a general methodology that is appropriate to the validation of bioeconomic models. There are, however, certain specific issues of model-testing that require attention before proceeding to formal validation. Checking the mathematical and logical correctness of the simulation model against the design criteria on which it was founded is an essential part of the building process without which formal validation is a pointless procedure. In a restricted sense, this procedure rightly may be considered to be verification.

Some authors have treated the terms 'validation' and 'verification' as synonyms (Naylor, Balintfy, Burdick & Chu, 1966; Naylor & Finger, 1967; Schrank & Holt, 1967) so that the terminology we have adopted is widely, but not universally, accepted. Others, however, have considered verification to be the process of ensuring that a model behaves as the experimenter intends it to, and validation to be the testing for agreement between model-behaviour and behaviour of the real system. Mihram (1972) defined verification as 'that stage of a model's development during which the model's responses are compared with those which would be anticipated to appear if indeed the model's structure were programmed as intended'. Mihram's (1972) definition supported that of Hermann (1967) and our practical interpretation of verification is that it is a process in model-development whereby the computer program of the model is checked for logical consistency.

Testing Models Against Design Criteria (Verification)

Comprehensive testing of the operation of any reasonably complex model is a formidable task. Usually it is prudent to begin this process by paying attention to individual modules before attacking the whole model. Nevertheless the approach is identical; we are searching the model for internal consistency. This is achieved by running the model with given input data and in prescribed exogenous conditions. Output from the model is generated so that its functioning can best be assessed. We recommend that this is always done without stochastic elements being included; that is to say the stochasticity is reduced usually to mean values. This makes the work of assessment more straightforward. It will still be an extensive task and gains can be made if part of it can be carried out by the model itself. The methods used to evaluate and correct a model include two important sets of techniques known as 'antibugging' and 'debugging'.

The terms antibugging and debugging are from the jargon of computer programming. A program with a bug is one which fails to behave in a consistent manner; it may terminate abnormally or produce obviously wrong answers either on every run or sporadically. Such a bug will be the result of some flaw in the program and one of the major costs in simulation modelling is in the detection and elimination of bugs. The process of tracing the cause of, and removing, bugs is known as debugging. Antibugging is the prevention—as opposed to the cure—of bugs and consists of techniques which make the occurrence of bugs less likely and reduce the time required to trace any remaining bugs.

ANTIBUGGING TECHNIQUES

Refer back to the final section of Chapter 3. There we were concerned with standards for computer programming which would help ensure the computer model was a proper representation of the model-design we had established. The guidelines we derived involved the application of antibugging techniques (although this is not explicitly stated in Chapter 3). As antibugging is a most valuable time and effort saver in model-building, the various techniques will be reviewed briefly below:

1. Incorrect data are one of the most common causes of bugs in simulation models. Data may be incorrect or may be incorrectly entered into the model. Therefore, as far as possible, all data input to a model should be checked for validity.
   (a) Incorrect or dubious data should generate an error or warning message which is readily seen and understood by the model-user. In any case, the facility for printing out input data in readable format should be built into the model.
   (b) Where card data are used, a consistent layout of data on cards should be used to reduce the chance of punching errors. It is also useful to make the card deck as visually readable as possible. Cards can be colour-coded into groups or identified by alphabetic or numeric code in a particular field on the card.
   (c) Simulation models used interactively should produce clear concise instructions on how to enter data and should permit the user to recover quickly and simply from any error in data input.
   (d) Where disk or tape data files are used, these should be carefully checked. The file may have been modified since the last simulation run either without the user's knowledge or without his recollection. One simple measure is to reserve one record on the

file (usually the first record) for identification purposes. Data held on this record could include

i. date of last use of file;
ii. date of last update;
iii. number of last update;
iv. identification of the last users to make an update and to access the file.

This information can then be printed out and can be checked either visually by the user or, in some cases, automatically by the model.

2. Maintenance of documentation is essential in efficient modelling. Refer back to Chapter 3 for detailed discussion on this issue.
3. Avoid multiple switches and program branches as far as possible. Each switch and branch point in the model is a potential source of bugs. Therefore the need for each switch or branch should be carefully evaluated and, where they are necessary, a decision table to examine all possible combinations should be drawn up as outlined in Chapter 3.
4. Many language compilers automatically check array subscripting and variable typing but some do not. Where this is not done by the compiler, incorrectly subscripted variables or variables which are intentionally, say, 'real' in one part of the model and 'integer' in another, can cause serious and unpredictable bugs.
5. A modular and systematic model-structure not only assists in conceptualising the model but also clearly is of assistance in isolating autonomous areas for verification.

Antibugging is really little more than a common-sense and systematic approach to computer modelling. The objective is to eliminate, as far as possible, those occasions when the modeller's intentions and the model's performance diverge due to errors in data or logic.

DEBUGGING TECHNIQUES

Invariably, a simulation model will contain bugs. These may be quite numerous the first few times the model is run but a model which has been properly antibugged should be relatively easy to free of these initial teething problems. Most bugs can be traced and eliminated relatively easily with a little careful and organised thinking. Bugs are mistakes and the more we learn about the kind of mistakes we are most likely to make, the easier tracing and eliminating bugs becomes. One approach to debugging is the shotgun approach where, for example, the model is made to print out the status of almost every variable at frequent intervals during the simulation run in the hope that the analysis of the remitting printout will indicate the cause of the problem. The result usually is that the model-builder finds himself deluged with more output than he can possibly interpret. This method is wasteful, costly and inefficient and we have developed the following guidelines to help in designing an effective debugging strategy.

1. Experiment with the bug. Determine whether the bug repeats itself and under what circumstances. A bug can mysteriously appear in a program simply due to an update in the computer operating system. If the model is used infrequently or has not been used for a period, this may be a possible cause. Write a small program to test the suspected cause of the bug and to see whether the bug can be reproduced in this program. If it cannot be reproduced the problem still requires analysis before further experimentation.

2. Remember your past mistakes. Most people have a pattern of thinking and very often the same basic mistake will turn up in a number of disguises. The modeller should keep a record of his debugging progress which may well indicate some pattern of errors. This can be used to eliminate other related errors and is invaluable for antibugging of future programs.

3. Trace the bug methodically and thoroughly. Try and determine which parts of the model have appeared to work satisfactorily and then check to see if they are indeed functioning as intended. Then work from the known to the unknown until the bug appears. Very often it will be a matter of 'zeroing-in' on the offending statement or program area.

4. Start with simple explanations and maintain an open mind. Most bugs are human mistakes and very often are due to quite simple causes. The modeller who made the mistake originally will have to alter his pattern of thinking to spot the error. Carefully examine all logic and assumptions. Yourdon (1975) provides a useful analogy between debugging and detective work. He suggests following the example of Agatha Christie's M. Hercule Poirot—avoid bloodhounds and massive dragnets, use the 'little gray cells'. Sit down and think, talk the problem over with colleagues or, in the last resort, just forget about it until tomorrow and do something else. The subconscious mind is often more flexible than the conscious and in many cases the bug can be quickly found after a break from the problem.

Debugging can be a frustrating and time-consuming process. However, if a model is well designed and documented, the likelihood is that the model-builder will soon become proficient at the detection and elimination of bugs. On the other hand, if the process of debugging seems endless, the model-builder may be well advised to return to the design stage of model-building and reconstruct his model in a more satisfactory manner. Models often grow by accretion; as they are used, new facilities are added which were not envisaged in the original design. The authors of this text have, on several occasions, decided that it was quicker and easier in the long run to redesign and build an existing model rather than attach and debug a new model-feature to an overburdened model-structure.

Validation of Simulation Models

We have divided the model-evaluation process into the model-testing (or verification) phase described above and a validation phase during which the model is assessed in relation to its prescribed use. On the issue of validation, three fundamental positions have been presented in the literature (Naylor, Balintfy, Burdick & Chu, 1966):

1. rationalism, a point of view which holds that a model is simply a system of logical deductions from a set of premises, which may or may not be open to empirical verification. Therefore, if these basic premises are accepted, then the validity of the model must also be accepted;
2. empiricism, which, in direct opposite to rationalism, holds that unless relationships and assumptions used in a model can be independently verified, then the model cannot be considered valid;
3. positivism, which accepts the validity of a model if it is capable of accurate prediction, regardless of the internal structure and underlying logic of the model.

Shannon (1975) considered the positivist approach to be that of the absolute pragmatists, but, when we consider the resource-demanding nature of model-building, this approach has appeal. Positivism supports the contention that costly simulation models should not remain 'white elephants' because they are deemed logically unsound but rather should be evaluated on their ability to fulfil the intended objective. Schrank & Holt (1967), for example, propose that the only criterion for model-validity should be its usefulness. The same authors argue that, by doing so, the emphasis is shifted from whether the model is a conception of truth or falsity to whether its errors render it too weak to serve the intended purpose.

The problem is, of course, that we will want to extend the model beyond the circumstances in which it has been tested when it comes to application. We may be nervous about doing this without more secure testing than positivism demands.

Naylor & Finger (1967) proposed another approach to model-validation which incorporates all the three views discussed above. In this multistage procedure (which Shannon, 1975, also calls a utilitarian approach) where it is not possible to subject all the structures in a model to formal empirical assessment, the testing of a selected number is considered sufficient. This again underlines the value of a modular approach so that individual modules can be separately tested prior to evaluating the model as a whole. Such a procedure attaches equal weight to the validity of hypotheses used for model-structure and the predictive capabilities of the model.

Much of the concern over validation of simulation models has come from economists and management scientists (Naylor & Finger, 1967; Schrank & Holt, 1967; Van Horn, 1971; Hermann, 1967) and has arisen partly due to the inability of these groups of researchers to obtain real data for testing their models. Although knowledge on biological systems is by no means complete in all cases, in general, biological-system researchers will find it easier to measure the input and output characteristics of the systems in which they are interested. This point will be brought out in the appendix to this chapter where the validation of the barley-leaf-rust model is discussed.

Validation Procedures

The validation procedure involves comparing the performance of the model either against recorded data for the system or against a subjective judgement of what the output should be, given a broad understanding of the system or type of system which the model represents. Clearly, in the first case, the model must be set up to conform to the exact circumstances of both controllable and non-controllable exogenous variables under which the real system was observed. Stochastic parameters in the model are therefore normally replaced by historical data on, for example, climate or prices. The model will be run in the exact time series of exogenous data for which we have records of the performance of the real system. Output parameters from the model over simulated time may be compared with measured output from the real system. The uncertainty surrounding relationships in most models will still be represented by error terms associated with the accepted functional form of the relationship. The result will be that the output parameter values will vary each time the model is run even though the exogenous stochastic parameters have been reduced exactly to recorded data. The time series of output from one run of the model is therefore one sample from a population. Similarly, the recorded measures of real-system performance are specific to the time and place characteristics of the system. These data too may therefore be considered as but a sample from a population. These considerations underlie the procedures of statistical analysis for validation.

In the situation where a subjective appraisal is the basis of validation, it is essential that all conditions of input to the model are detailed over simulated time. The fact that one run of the model produces but one sample output from the population should be made clear to the assessors, and where possible the output time series from a number of runs should be provided.

STATISTICAL TESTING

The fact that statistical testing is considered implies that some real data for comparison with model-output are available. Anderson (1974) identifies four sources of data for validation:

1. Historical data already used for model-building.
2. Historical data not used in model-building.
3. Historical data collected since the model was completed.
4. Data specifically generated or collected for validation.

Of the four sources of data, type 1 is not acceptable for validation for obvious reasons, though it should be noted that it is essential for verification. As far as type 2 data source is concerned, its existence implies a specific decision to exclude from model-building some portion of available information. Often this will not be possible or desirable. The third and fourth data sources are most suitable but often are not available. For some biological systems, new data may be quite readily generated: this was so in the case of the barley-leaf-rust system as illustrated in the appendix in this chapter.

A first step which we always recommend in validation is drawing out the data in a time series with real-system recorded performance and model-output on the same graph. The time series for exogenous variable data will be identical but, as explained above, error terms within the model will generate a different sample of output each time the model is run. In this case, it is suggested that the means of the model-output series be used. An

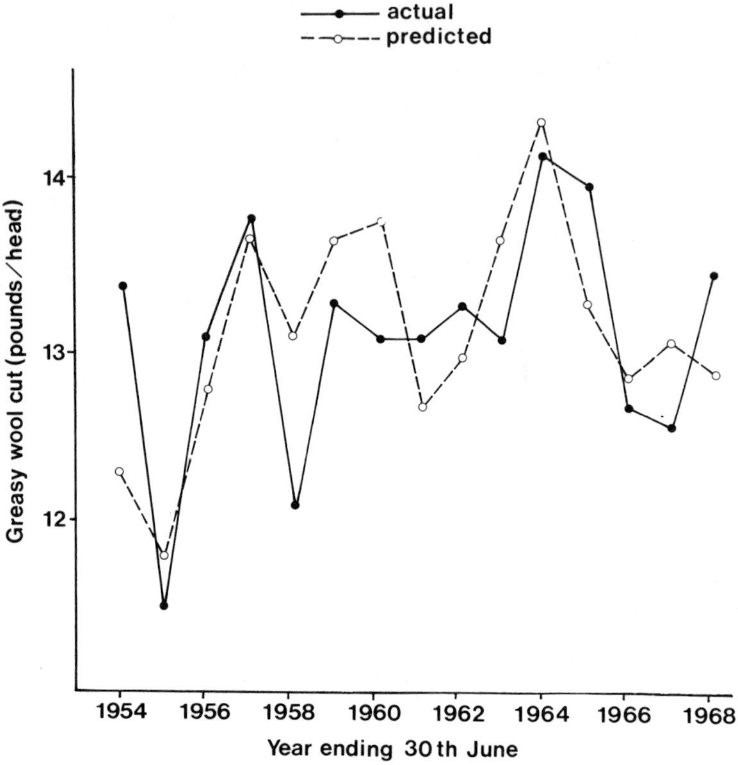

Fig. 5.1. Graphical display of model-output and real-system records for greasy wool clip from a farm in the arid region of Australia. (After Chudleigh, 1971.)

example of the plot-out of the two sets of data are given in Fig. 5.1 from the work of Chudleigh (1971).

The model involved predicted sheep numbers and wool clip per sheep, as well as associated financial parameters for a representative farm in the arid region of Australia. Greasy wool clip per head actually taken on the farm in each year from 1954 to 1968 is plotted alongside mean model-output over the same years. This procedure is simple and provides a 'feel' for the situation and a good basis for further analysis.

Comprehensive reviews of statistical tests suitable for validation have been provided by a number of authors (for example, Naylor & Finger, 1967; Mihram, 1972; Kleijnen, 1974). From these, two general approaches are now briefly described.

1. Simple regression analysis is performed between mean model-outputs and real-system measures as paired observations. A linear regression is fitted and a perfect model would provide a line passing through the origin with a slope of 1, i.e. a 1:1 correspondence. The test is to determine whether the intercept is significantly different from zero and the slope is significantly different from unity.

A model of the lucerne crop was capable of predicting yields under alternative management treatments and in relation to climate and soil type (Richie, Dent and Blackie, 1978). Trial data were available which measured lucerne dry-matter yields over 78 sequential cuts over a number of years. The model was set up for a single run in the same sequence of climate data (historical data input) and was subject to precisely the same cutting management. The regression equation derived from these data was

$$\begin{aligned} Y = &-43{\cdot}72 + 0{\cdot}95X \\ &(309{\cdot}1) \quad (0{\cdot}121) \end{aligned}$$

$$R^2 = 0{\cdot}44; \text{ Standard Error Estimate } 928{\cdot}061$$

The null hypothesis that the slope coefficient is not significantly different from zero is tested with $n - 2$ degrees of freedom by the Student t-test. The value of t for the slope coefficient is

$$\frac{0{\cdot}95 - 1{\cdot}0}{0{\cdot}121} = -0{\cdot}381$$

This case is not significantly different from 1. Similarly the intercept is not significantly different from zero with a value for t of

$$\frac{-43{\cdot}72 - 0}{309{\cdot}1} = -0{\cdot}141$$

Perhaps a more relevant null hypothesis is that the intercept coefficient and the slope coefficient simultaneously are not different from zero and unity respectively. The appropriate statistic in this case is the F statistic:

$$F = \frac{\{n - 2\}\{n(a)^2 + 2n\bar{X}(a)(b - 1) + \Sigma X_i{}^2(b - 1)^2\}}{2n(S)^2}$$

where X_i are the individual model-output values; $\bar{X}$ is the mean of model observations; and S is the standard error of the estimate. The computed value of F can then be compared with tabulated values at 2 and $n - 2$ degrees of freedom.

A word of caution has been sounded by Aigner (1972) with regard to these simple tests. He argues that, for the case of a stochastic simulation, the regression of actual on model-generated values (or vice versa) should, in fact, not be expected to be homogeneous with unit slope. On the other hand, Aigner (1972) and Howrey & Kelejian (1969) are in general agreement with this test for deterministic models. The above authorities, however, can suggest no clear alternative for stochastic simulation and draw attention to the need for further research on this problem.

2. There are a number of specific tests to examine the hypothesis that the simulated and the real recorded-system outputs have the same parent distribution. Usually these are referred to as goodness-of-fit tests. The null hypothesis is one of sameness while the alternative hypothesis is that the two sets of distributions do not have the same parent distribution. Should the null hypothesis be rejected, the alternative is accepted with a certain probability of a Type I error—the error of rejecting a true null hypothesis. It is the Type I error which is of obvious relevance in conventional statistical analysis where, for example, a scientist may wish to show that a new crop variety yields significantly more than an existing variety. Thus the test is being used to establish a possible difference between the two sample populations. Scientific literature abounds with statements along the lines that treatment A is different from treatment B at the 5 per cent level of significance. In this case we are assumed to be 95 per cent confident that the sample distributions from treatment A and treatment B do not come from the same parent population, i.e. we reject the null hypothesis with 95 per cent confidence.

In validation, the modeller wishes to establish the sameness of model output as historical record. It is the probability of a Type II error (the error of accepting that these are the same when, in fact, they are not) which should be of major concern (Orcutt, Greenburger, Korbel & Rivlin, 1961; Stash, 1969). However, the null hypothesis that the distribution of samples of model-output and the observed real-system distribution are from different populations is too vague to secure a knowledge of a Type II error. Notice though that the probability of a Type II error can be reduced by increasing the chance of a Type I error. Hence the model-builder concerned with validation may be seen to serve his interests most effectively by increasing the confidence level associated with Type I error (Greig, 1978a).

If under conventional statistical analysis model-output is considered to be significantly different from historical record with a high degree of confidence, Greig (1978a) has suggested that there are two courses of action open to the modeller. The first is to rework the model until it is acceptable—i.e. no highly significant difference is detected. The second is simply to cease the whole modelling effort altogether, though at this stage this option seems unlikely to be exercised particularly since some benefits to modelling initially have been envisaged and the modelling effort to date is a heavy sunk cost. To forsake anticipated benefits for some marginal modelling effort is likely to incur substantial opportunity cost. The direct costs associated with Type I error, therefore, are likely to be some additional modelling effort in an attempt to secure acceptance of the model.

Greig (1978a) goes on to assert that the cost of making a Type II error is likely to be much more substantial. Here we are accepting a model as valid when in fact it is inadequate: the costs are those of making wrong decisions using the invalid model. The magnitude of this cost will not clearly be known in advance even if the decision maker intended to rely only on the model to influence his choice. It could, however, be extensive, and eventually be associated with the long-term credibility of modelling as an aid to decision making.

The interpretation of results from statistical tests should be critically determined in the light of the above discussion. Even though some of the tests are relatively simple to apply and the results may appear clean-cut, the modeller should beware lest he be misled by the superficial objectivity. Statistical tests involving the assessment of Type I error can only be a broad guide and should be applied in this light. Common tests include the chi-square test and the Smirnov test and readers will find details of these and other suitable techniques in the text by Conover (1971). The appendix to this chapter contains an example of the application of the Smirnov test with data from the barley-leaf-rust model.

SUBJECTIVE ASSESSMENT

Statistical testing of a model has an aura of objectivity and yet, as we have seen, major problems of interpretation remain. As a result some authorities, including Greig (1978b), have advocated less 'objective' approaches exemplified by simple graphical display of model-output in relation to the controllable and uncontrollable exogenous variables and where possible relative to real-system output as in Fig. 5.1. The subjective interpretation and assessment of graphic display by the model-building team and by others judged expert in the area are in fact the basis of the Turing-type test. (Mitroff, 1969). The basis of these tests is simply to present information on model-performance to an authority or a panel of experts with a request to comment on the reality of the data. It will be noticed immediately that such tests may be carried out with or without recorded historical data about the real system and therefore can apply to models of

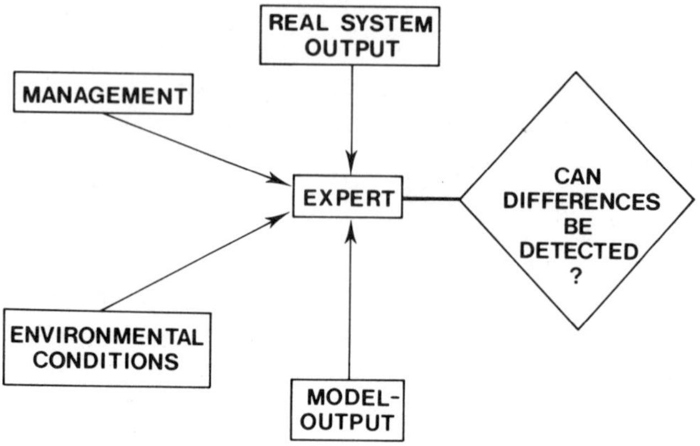

Fig. 5.2. Subjective assessment procedure when real-system data are available.

systems not yet in existence or for which no recorded performance data are available.

1. Comparing real and model-output by a Turing-type test can be achieved in the manner suggested in Fig. 5.2.

In this case the model is run in precisely the same set of conditions as the real system experienced during the recording period. This, of course, means supplying historical time-series data for uncontrollable exogenous variables and for the imposed management so that the model-performance is directly comparable with the real system. Several runs of the model are desirable to provide a sample of performance levels. These data, together with the monitored real-system performance, are presented to the expert with a request to determine whether or not he is able to detect important differences and whether he can isolate which output data are from the real system and which from the model. The method of presentation is important and we feel that the best approach is graphical display of major output variables against time with the levels of the main exogenous variables also graphed. Care should be taken so that the selected scales do not distort the picture and in this regard it is usually wise to provide tabular data alongside the graphs. The modeller must also decide whether a single expert, a sequence of experts or a panel of experts should assist him in the validation. Certainly the modeller should try to be aware of his own biases and avoid presenting data in such a way as to support them. Very often it will be desirable for this Turing-type testing to proceed over a protracted period of time and possibly at various stages of completion of the model. Undoubtedly the testing of individual modules in this way is beneficial particularly since specialists can be called for specific areas of the model.

2. Assessing model-performance when no recorded information about a real system exists may follow the outline of Fig. 5.3.

Here, the modeller provides the expert with input and output time series for the model. The model may be run in stochastic mode so long as the input levels are recorded. Conversely, it may be felt that a time series of historical

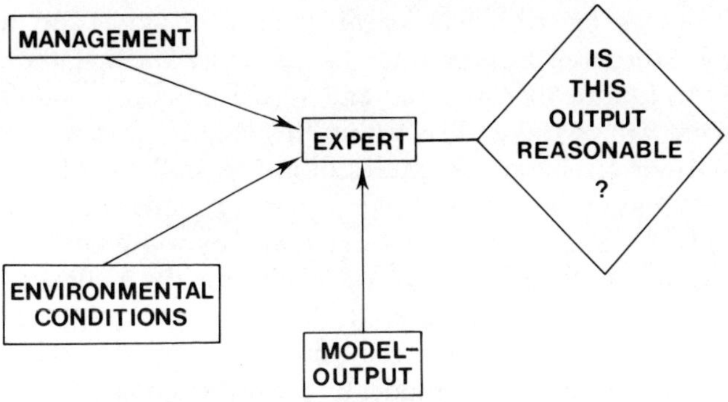

Fig. 5.3. *Subjective assessment procedure with no real-system data.*

input data for the exogenous variables may prove more useful in the early stages of assessment at least so that the expert might be presented with more confined and more familiar circumstances. The expert is then requested to comment on the input–output series from the model with an eye to the acceptability of relationships so formed.

Again, time-series graphic display of input and output seem to be the most favoured method of presenting the data. Because the expert is being called on to make extensive subjective assessment, perhaps about systems which do not yet exist, it may well be that he should be given the opportunity to request output related to management treatment which he imposes. This facility will help him to gain a feel of how the model responds to changes in management variables under varying environmental conditions. Apart from this important point, comments made about Turing-type testing for comparison of model and real-system data apply to testing where no real-system data are available.

A Long-term View of Model-evaluation

The fact that one or a series of validation tests indicate that the model is performing adequately must not be considered the end of validation. The model has been assessed either against one sample of the performance of the real system or against a limited number of evaluations by experts. Certainly our confidence in the model will have been enhanced but for certain prescribed circumstances: different management or different environmental conditions may see the model performing less satisfactorily. So it is that validation is a continuing process during which confidence in the model (hopefully!) steadily increases through a succession of formal and informal tests. During this time we will be learning to use the model and, in using it, probably will discover the need for enhancements. The updated version of the model will be further assessed and so the process of validation and improvement will continue. It is within this long-term view of validation that the various tests and procedures of assessment should be applied.

Model-evaluation is then an integral part of model-building and use. Models must be built in the knowledge that they will be modified during evaluation and use. Validation is a crucial phase in the modelling work but it is not a 'one-shot' procedure. Rather it is a continuing series of assessments which continue into the commissioning stage of the model's life. Critical assessment of the model by the users is bound to occur in the early stages of application. Involvement of the user of the model, whether a commercial manager or a research scientist, in the validation process is highly desirable and in the long run saves time. The requirement of the user for assessment will frequently govern the type of testing most appropriate for any particular model.

Appendix

VALIDATION OF THE BARLEY-LEAF-RUST MODEL

Validation data for the barley-leaf-rust model were collected over two cropping seasons. A series of trial fields were planted to barley, and epidemics started in these fields at various times during the growing season. Data were collected on the actual progress of the various epidemics and these were compared with results from simulated epidemics predicted by the model under identical environmental conditions.

Graphical display and assessment

A visual comparison was made between the progress curves of field epidemics from five sites and the corresponding curves produced by the model. Two examples of graphs produced of observed and simulated disease curves are shown in Figs. A5.1 and A5.2. Figure A5.1 illustrates an epidemic of short duration and with a low terminal severity while Fig. A5.2 shows the curve generated by a long duration, high terminal severity epidemic. Graphical comparisons of the various other epidemics investigated showed that the simulated data were similar to observed disease data in all cases.

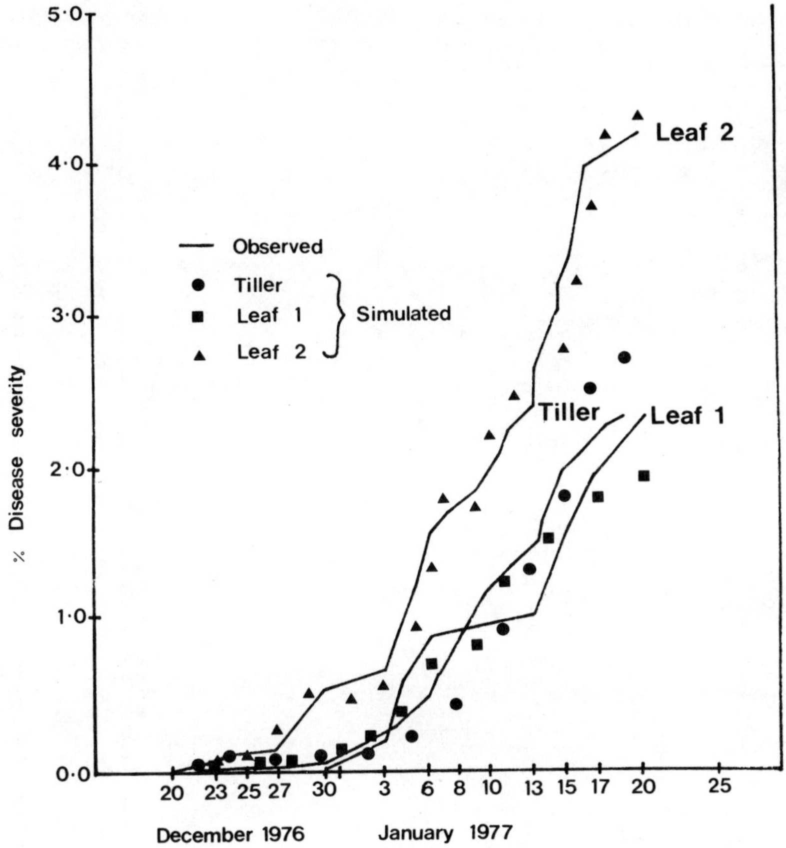

Fig. A5.1. Observed and simulated output for an epidemic of short duration and low terminal severity. (After Teng, Blackie & Close, 1978.)

All epidemics generated by the model showed the sigmoid curvature of barley-leaf-rust epidemics. In the epidemic illustrated in Fig. A5.1, where epidemic onset was relatively late in the crop's growth and disease severity did not exceed 5 per cent of leaf area, model-output was still seen to agree with field data showing that the model performed satisfactorily over the whole range of expected epidemic situations.

Scatter diagrams relating observed disease values to model-generated values in all cases showed points lying close to a 45° straight line: examples from two fields are shown in Fig. A5.3.

In all, data from 15 imposed epidemics were made available over a 2-year period against which the performance of the model was checked. The 15 epidemics were representative of the wide range of naturally occurring disease outbreaks and the

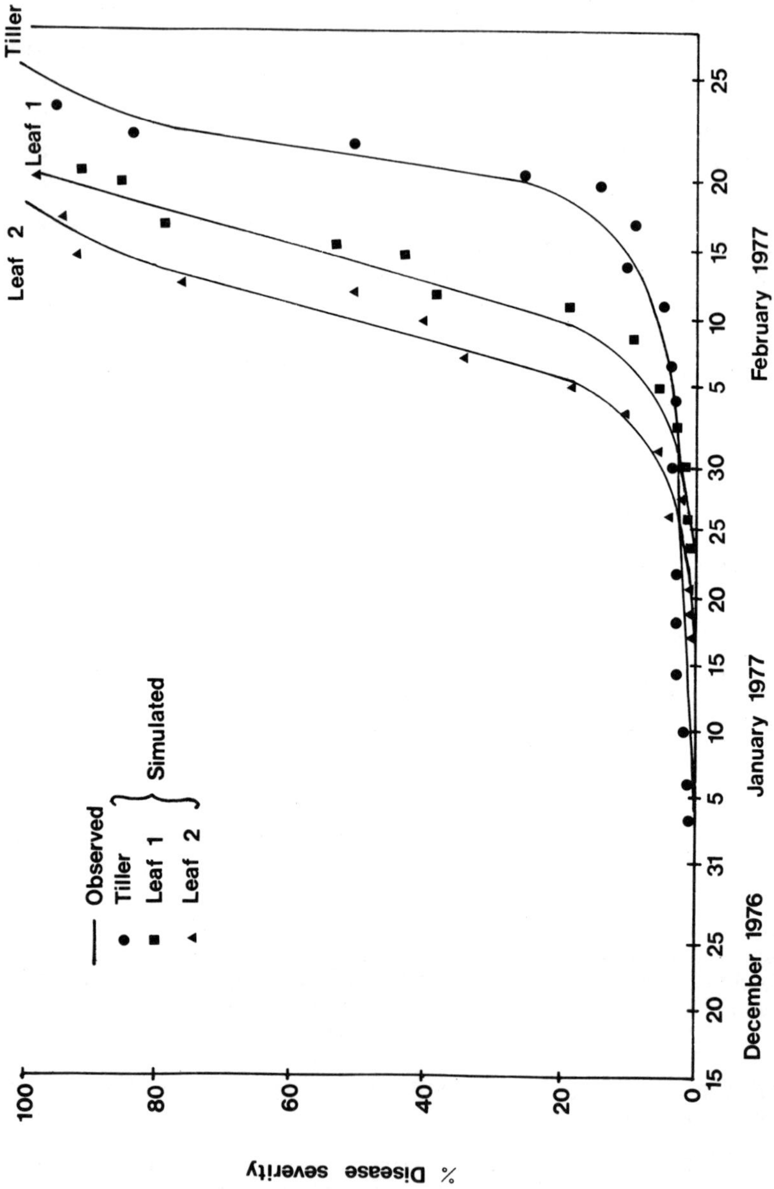

Fig. A5.2. *Observed and simulated output for an epidemic of a long duration and high terminal severity.* (*After Teng, Blackie & Close*, 1978.)

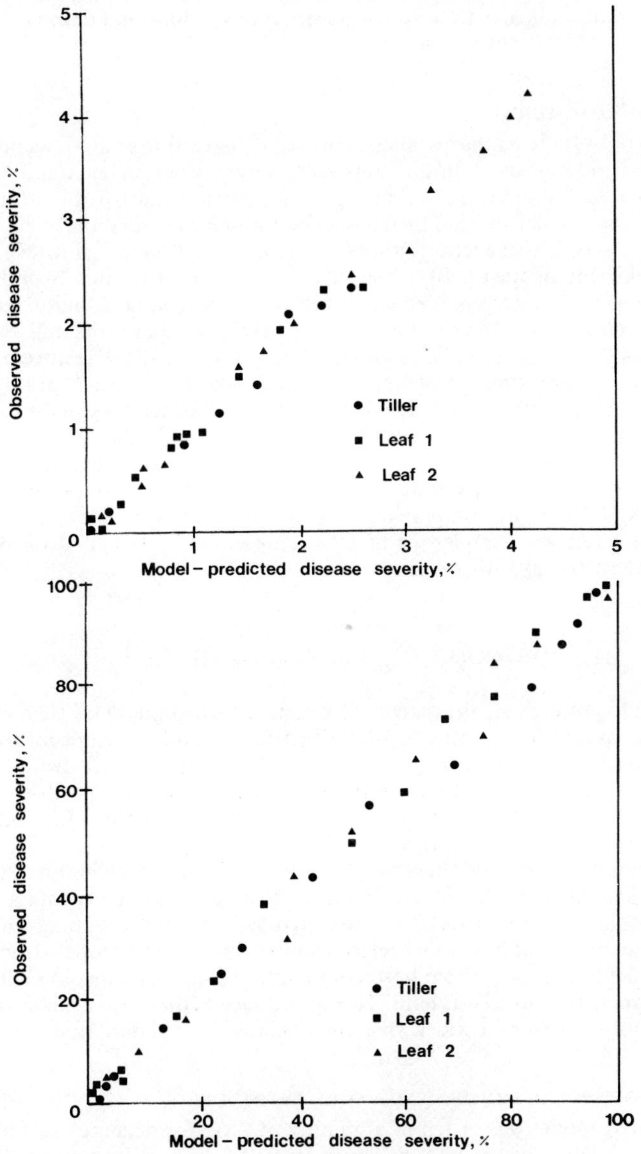

Fig. A5.3. *Scatter diagram relating observed terminal severity to model-predictions.* (*After Teng, Blackie & Close*, 1978.)

fact that these could be established from field trials to specifically produce validation data indicates that, for some situations, additional data can in fact be produced quickly enough to serve model-assessment procedures.

Statistical assessment

The field trial data from the five sites were used in regression analysis against model-output under identical conditions. Separate analyses were carried out for each field on the tiller, leaf 1 and leaf 2 over the 15 different epidemics. Table A5.1 provides the regression coefficients and their associated standard errors, the value of R^2, the value of t for each coefficient (testing the intercept coefficient against zero and the slope coefficient against unity) and the F value for the null hypothesis that simultaneously the intercept is equal to zero and slope is equal to unity. High values for R_2 are a feature of the results as well as general support for the null hypotheses: in nine cases the null hypothesis on both the t-test and the F-test were upheld.

Goodness-of-fit testing proceeded using the Smirnov test, details of which can be found in Conover (1971). For each field and for each of the three leaves separately the distribution of per cent severity of infection over the 15 epidemics was established. Similar data from the model-results provided an analogous distribution. These density functions were converted to cumulative probability functions (CDF) before comparison using the Smirnov test. The test assumes two independent random samples with CDFs represented by F(x) and G(x). Using a two-sided test the hypotheses are

$$H_0\colon F(x) = G(x) \quad \text{for all } x \text{ from } -\infty \text{ to } +\infty$$
$$H_1\colon F(x) \neq G(x) \quad \text{for at least one value of } x$$

If the null hypothesis H_0 is accepted at a certain probability level then we can say that model-output for the situation described does not differ significantly from field data at this level. The test statistic is T which is the greatest vertical distance between the two distributions and is compared with tabulated values at a specified confidence level. Results from the Smirnov test are presented in Table A5.2 and the null hypothesis is accepted in all cases.

The validation described, however, must be seen in perspective. First, the model had only simulated epidemics occurring on five fields located within a radius of 1 km. Secondly, validation data were collected over only two cropping seasons, and thirdly, the biological functional relationships built into the model-structure were for a 'virulent-race-susceptible host' combination found at Lincoln College, New Zealand where the model was built. The significance of these three factors cannot be judged until data from a wider area are available for validation.

VALIDATION OF THE PIG-HERD MANAGEMENT MODEL

The pig-herd model was tested in a number of ways for accuracy of future event predictions against historical time series recorded on trial farms. Recall that the model was constructed to include two submodels—one simulating the breeding herd and the other the fattening herd. Each submodel may be operated as an independent unit and, therefore, required separate validation.

Table A5.1. Results of linear regression analysis of observed disease severity on model-predicted disease severity
(After Teng, Blackie & Close, 1978.)

Field	*Leaf*	*a*	*Standard error (a)*	t_a	*b*	*Standard error (b)*	t_{1-b}	*F*	R^2
1	Tiller	0·146 0	0·098 2	1·487	1·000 0	0·006 3	—	0·18	0·999
	Leaf 1	0·135 4	0·134 7	1·005	1·002 8	0·005 1	0·549	0·71	0·999
	Leaf 2	−0·185 2	0·058 5	3·166**	1·007 6	0·003 9	1·949	10·99**	0·999
2	Tiller	−0·007 9	0·117 9	0·067	0·943 0	0·012 6	4·524**	2 004·13**	0·999
	Leaf 1	0·053 3	0·176 0	0·303	0·931 4	0·122 4	0·560	6·41**	0·980
	Leaf 2	−0·007 5	0·001 1	6·818**	0·979 0	0·014 6	1·438	2·47	0·998
3	Tiller	−0·134 1	0·007 1	18·887**	1·007 3	0·002 7	2·704*	7·64**	1·000
	Leaf 1	0·207 8	0·106 5	1·951	1·010 4	0·007 0	1·486	4·71	0·999
	Leaf 2	−0·383 1	0·055 8	6·866**	1·002 6	0·005 0	0·520	8·43**	0·999
4	Tiller	−0·120 9	0·262 7	0·460	1·014 4	0·017 1	0·842	0·09	0·999
	Leaf 1	−0·141 8	0·214 3	0·662	0·979 3	0·009 1	2·275	0·38	0·998
	Leaf 2	0·082 8	0·079 0	1·048	1·009 7	0·004 9	1·980	0·20	0·999
5	Tiller	−0·001 5	0·017 8	0·084	1·014 4	0·004 4	3·273**	21·70**	0·999
	Leaf 1	0·135 5	0·273 3	0·496	0·979 3	0·012 6	1·643	1·42	0·999
	Leaf 2	−0·202 5	0·112 9	1·794	1·009 7	0·006 9	1·406	2·34	0·999

* indicates difference at the 95 per cent level of confidence.
** indicates difference at the 99 per cent level of confidence.

Table A5.2. Results of Smirnov test on disease severity
(After Teng, Blackie & Close, 1978.)

Field	*Leaf*	*T*	*Two-tailed quantiles*	
			$W_{0.95}$	$W_{0.99}$
1	Tiller	0·1066	0·4737	0·5678
	Leaf 1	0·1212	0·6113	0·7326
	Leaf 2	0·1636	0·5942	0·7122
2	Tiller	0·1103	0·4737	0·5678
	Leaf 1	0·1786	0·7039	0·8436
	Leaf 2	0·1515	0·8786	1·0531
3	Tiller	0·1538	0·5444	0·6525
	Leaf 1	0·2222	0·6608	0·7926
	Leaf 2	0·0833	0·5677	0·6804
4	Tiller	0·1333	0·5053	0·6057
	Leaf 1	0·2500	0·7039	0·8436
	Leaf 2	0·1727	0·5942	0·7122
5	Tiller	0·1176	0·4737	0·5678
	Leaf 1	0·2000	0·6249	0·7489
	Leaf 2	0·1889	0·6249	0·7489

The criterion of performance of the breeding unit was taken as the number of farrowings and Fig A5.4 indicates the mean percentage error of forecast over all sows and through time for two separate farms. Records of farrowings for each farm

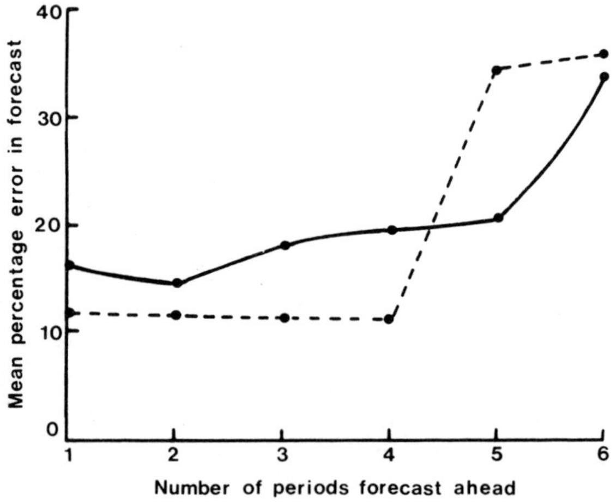

Fig. A5.4. Farrowing forecast errors for two trial farms. (After Blackie and Dent, 1974.)

were kept and actual conditions at the start of the forecast period were input to the model. Model-forecasts of farrowings were compared in each month with the recorded number of farrowings. A forecast error is incurred whenever the number of farrowings did not coincide with the number forecast for that particular month. The variability in biological systems makes event forecasting somewhat unpredictable and a degree of error is inevitable. In fact, if a farrowing was forecast to occur on the last day of a month and actually occurred on the first day of the next month an error is indicated. In the absolute case this could mean an extremely small error one side or the other of midnight. The test is therefore a severe one. In addition, the model is obviously constrained to predictions of whole numbers of farrowings; in Fig. A5.4 a difference of a single farrowing between the period forecast for an event and its actual occurrence in another period incurs an error level of approximately 5 per cent. From Fig. A5.4, it is apparent that, for the farms involved, the quality of model for event prediction is approximately constant for 4–5 periods ahead (roughly equivalent to one production cycle) but falls off after this time. The indicated increase in forecast sequence error over longer periods is to be expected; management decisions may be altered or have their timing changed due to influences outwith the body of the model. In the shorter term, the apparently high error rate can be explained in terms of the time period and integer effects expounded above.

When the total number of farrowings over the whole 6-month period was examined, a much lower level of error was involved. The mean percentage error for total farrowings occurring over a 6-month prediction period for both farms was 1·59 per cent. This error rose to 7·36 per cent for a prediction period of 1 year.

Validation of the fattening sector was more subjective due to the peculiar features of this part of the model. Any failure of the predictions of the selected criterion (number of pigs sold) to correspond with reality can be attributed in part to the following causes:

1. Initial failure to classify fattening pigs exactly into their correct weight bands. (Data input to the model includes *estimated* numbers of pigs in various weight categories. Practical considerations rarely permit actual weighing of fattening pigs.)
2. Failure to achieve the mean growth rate implied in the input data.
3. Changes in selling policy and/or weights not notified in the input data. (Pigs may be sold at a number of different slaughter weights and many producers sell pigs at more than one of these weights.)

Sensitivity analysis previously indicated that model-predictions were not seriously affected by minor errors in assigning fattening pigs to the various initial weight categories. Experienced pigmen can be expected to estimate weights of fattening pigs within about 10 kg (at the most) of actual weights. This degree of error affected model-predictions but within accepted tolerances. Failure to achieve the growth rate implied in the input data or changes in sales policy were the chief sources of error in model-predictions. The results from all farms where the model was tested indicated that pig-growth predictions were in line with the information provided by the managers. The major difficulty with respect to validation remained the discrepancy between the selling policy which actually occurred and that stated by the manager in the setup data.

A change in validation procedure was therefore adopted which involved using a historical time series of information about the pig units including the managers actual selling policy. For each farm the selling policy over the period of validation was input as data along with an assessment of the distribution of pig weights at the start of the period. This procedure provided a method whereby any disturbance caused by management changes could be eliminated. Direct comparison of model-output and recorded data was now possible and the mean error for the total number of fattening pigs sold over a 12-month period and for all test farms was 1·29 per cent.

For this model, specific statistical testing was not adopted mainly because the users (farmers and farm advisers) found the type of testing described acceptable. Explaining the results of statistical testing to managers in general is often a less satisfactory approach than simple diagrammatic and tabular presentations. Validation should never be seen as an end in itself; the procedures adopted must depend on the intended use to be made of the model.

This appendix has reinforced some of the approaches presented in the chapter but it should be stated that the data presented here do not represent the complete validation of the two models. What has been described are some of the specific tests carried out, tests which improved confidence in the adequacy of the models. As explained in the chapter, confidence grows over time with testing and with use and no individual series test can be considered sufficient to complete the validation of a model.

References

Aigner, D. J. (1972). A note on verification of computer simulation models. *Man. Sci.*, **18,** 615–19.

Anderson, J. R. (1974). Unpublished manuscript.

Blackie, M. J. & Dent, J. B. (1974). The concept and application of skeleton models in farm business analysis and planning. *J. Agr. Econ.*, **25,** 165–73.

Chudleigh, P. D. (1971). Unpublished Ph.D. Thesis, University of New South Wales, Australia.

Conover, W. J. (1971). *Practical Nonparametric Statistics.* New York: Wiley.

Greig, I. D. (1978a). Unpublished Ph.D. Thesis, University of New England, Australia.

Greig, I. D. (1978b). On Statistical Testing and the Validation of Simulation Models (with editors *Agr. Systems*).

Hermann, C. (1967). Validation problems in games and simulation. *Behav. Sci.*, **12,** 216–30.

Howrey, E. P. & Kelejian, H. H. (1969). Simulation vs analytical solutions. *In:* Naylor, T. H. (ed), *The Design of Computer Simulation Experiments.* Durham, N. C.: Duke University Press.

Kleijnen, J. P. C. (1974). *Statistical Techniques in Simulation.* New York: Marcel Dekker.

Mihram, G. A. (1972). *Simulation: Statistical Foundations and Methodology.* New York: Academic Press.

Mitroff, I. I. (1969). Fundamental issues in the simulation of human behaviour. A case in the strategy of behavioural science. *Man. Sci., Application Ser.*, **15,** 635–49.

Naylor, T. H., Balintfy, J. L., Burdick, D. S. & Chu, K. (1966). *Computer Simulation Techniques*. New York: Wiley.

Naylor, T. H. & Finger, J. M. (1967). Verification of computer simulation models. *Man. Sci.*, **14,** 92–101.

Orcutt, G. H., Greenburger, M., Korbel, J. & Rivlin, A. M. (1961). *Microanalysis of Socio-economic Systems: A Simulation Study*, New York: Harper.

Ritchie, I. J., Dent, J. B. & Blackie, M. J. (1978). Irrigation management: an information system approach. *Agr. Systems*, **3,** 67–74.

Schrank, W. E. & Holt, C. C. (1967). Critique of verification of computer simulation models. *Man. Sci.*, **14,** 104–6.

Shannon, R. E. (1975). *Systems Simulation: The Art and the Science*. Englewood Cliffs: Prentice-Hall Inc.

Stash, S. F. (1969). Some aspects of multidimension and verification of simulations. Abstract in Naylor, T. H. (ed), *The Design of Computer Simulation Experiments*. Durham, N. C.: Duke University Press.

Teng, P. S., Blackie, M. J. & Close, R. C. (1978). Simulation of barley leaf rust: structure and validation of BARSIM, I (with editors, *Ag. Systems*).

Van Horn, R. L. (1971). Validation of Simulation Results. *Man. Sci.*, **17,** 247–58.

Yourdon, E. (1975). *Techniques of Program Structure and Design*. Englewood Cliffs: Prentice-Hall Inc.

6

Design of Simulation Experiments

Following the exhausting demands of constructing and testing the model, it is frequently the case that insufficient time and thought are given to experimentation with systems models. However, careful planning of simulation experiments is essential if the potential usefulness of a model is to be fully realised. Appropriately designed experiments can enhance understanding of the relationships and behaviour of the system in explanatory studies or can reveal how management might control the system so as best to achieve certain goals in prescriptive (normative) studies. However, the experimental techniques used may be very different from those in conventional (real-life) experimentation and, in this chapter, we will examine the design and analysis of computer simulation experiments.

It will be recalled that an essential reason for designing and building the simulation model was that information could not be derived from experimentation on the real system: either the system as such may not be in existence or experimentation may be too costly, time consuming or ineffective. Part of the cost of experimentation on the real system is the possibility of damaging or altering it in ways not foreseen prior to experimentation. This is particularly relevant in delicate ecological systems or in business and economic systems. On the other hand, provided the model has been satisfactorily validated, the response of the real system may be implied from the response of the model.

The traditional designs of experiments performed in the field, glasshouse or laboratory are suitable for computer experimentation and are, in fact, often used for this purpose. However, simulation experiments have important differences from real-world experiments; in particular

1. total 'control' over the experimental environment is possible by way of the exogenous variable inputs;
2. real time is compressed within the computer operation of a system so that time delays in experimentation are minimal;

3. experimental treatments are always sequentially explored unlike the usual simultaneous treatment evaluation in real life.

These features open the way to more efficient experimental-design procedures and allow a substantial increase in the number of factors that can be independently varied, while at the same time imposing greater responsibility on the researcher to select an appropriate design layout.

Factors, Treatments and Replicates

Before examining specific designs it is necessary to introduce certain terminology and concepts relating to agricultural experimentation in general. Regardless of whether the experimentation is carried out on a real system or a computer model, there will be certain variables we wish to adjust or set at different levels, and these are known as experimental *factors*. The factors may be qualitative, such as crop variety, type of pesticide, or whether crops are irrigated or grown under natural rainfall. They may also be quantitative, e.g. fertiliser application rate, amount of water per irrigation or ration quantity for livestock. Usually we will wish to vary more than one factor during an experiment. For example, in a fertiliser trial the application rates of nitrogen (N), phosphorus (P) and potassium (K) may be set at various levels. Each combination of factor levels is known as a *treatment*, e.g. N, P and K levels of 50, 10 and 10 kg/ha constitute one treatment while levels of 60, 5 and 20 kg/ha would be another. An *experiment* (real-world or with a simulation model) involves the evaluation of two or more factor combinations or treatments.

Of course, the performance of a system will depend not only on the particular treatments we choose to impose but also on a wide range of other factors. In general we may say that system performance is a function of a number of exogenous variables, some of which are under the control of the experimenter and some of which are not. For example, a researcher conducting a crop fertiliser trial may be able to choose the variety, plant population, row spacing and location of plots, and may exert a good deal of control over date of planting, weed population and pest and disease incidence. However, yield fluctuations not due to experimental treatments will still arise because of irregularities in soil fertility between plots, genetic differences in seed and other factors. As well, if the experiment were carried out at some other location, on a different soil type, with another variety or in another season, then the results could be quite different. In constructing an experiment, an endeavour is made to compare the various treatments imposed under, as near as possible, identical conditions. The concern, therefore, is to arrange the experiment so that the effects of the non-controllable sources of variation are as small as possible. Hence, an experiment is so situated that either the general environment is repeatable in other places or the results from the experiment are directly related to and used only for the particular region where it is carried out. When we experiment with laying hens, for example, we would record the temperature and humidity in the house and make a major effort to control these factors so that the environment can be repeated. In a crop experiment we would place all treatments spatially close together to reduce the effect of variation in soil conditions. By selecting biological material from similar sources (the same litter or the same strain of crop variety) we are again helping to reduce uncontrolled variation.

The fact that most applied biological experiments are replicated (i.e. each treatment is included in the design two or more times) suggests that there is only limited success in controlling the environment to eliminate unwanted sources of variation. Replication is necessary so that we are in a better position to draw inferences about the manner in which the population of the biological material involved behaves from the relatively small number of observations made in the experiment. The larger the sample in the experiment, the greater the probability that the sample averages will be close to the population (system) averages. Thus, the greater the number of replicates of each treatment the more we can be sure that the mean result of these replicates will be a good reflection of the way the system (on average) would react to the treatment. In order to compare various treatment effects satisfactorily, the number of replicates of each treatment may have to be large.

As we increase the number of replicates, it is implied that we approach a more accurate knowledge of how the system on average would react to the treatments imposed. Unfortunately, however, we have to make this statement a conditional one because it is only true relative to the general environmental circumstances in which the experiment was carried out. We may or may not be able to estimate the parameters of this environment and only rarely (as in the case of 'controlled-environment' experiments in intensive livestock houses or in the use of a phytotron) we will be able to control them. If we wish to know how the system will respond to the same treatments under different environmental conditions, the experiment must be repeated under these changed conditions. Considerations of this nature assume greater importance when the experiment is carried out over a number of periods (e.g. years), each with a particular set of environmental parameters. Experimental results from such a programme may have very limited application since they may be valid only for the particular sequence of periods and in the specific region where the experiment was executed.

Computer-based Experimentation

Following this brief review of experimentation in general, we can go on to look at experimentation with computer models in more detail. Assume that we are in the position of having a validated and operational model of a particular bioeconomic system, and that the objective is to determine the appropriate organisation of this system to optimise performance, e.g. to maximise output, yield or profits or to minimise costs.

Just as simulation models are made up of symbolic expressions, so it is convenient to characterise the relationships to be investigated experimentally in symbolic form:

$$Z = \mathrm{f}(X, Y)$$

where Z is a criterion of the systems performance (e.g. yield, net income), X represents the controllable factors, and Y represents the non-controllable factors. Each of X and Y may be a single variable or a vector containing a number of variables. The model is in essence an expression of the relationship between performance Z and the variables in X and Y. In fact, from the point of view of our experiments we may regard the model as simply a procedure (expressed in symbolic language and programmed for a computer) that will predict the level of Z for any given X and Y. The X vectors may be regarded as experimental treatments or management policies while the Y vectors are uncontrollable exogenous variables outside the model-boundary, under which the policies are to be evaluated. Each replicate of a given treatment utilises a different set of non-controllable factors or 'states of nature', i.e. a different set of values for the variables represented by Y. The procedures followed in a simulation experiment are shown in Fig. 6.1.

SINGLE-FACTOR EXPERIMENTS—DETERMINISTIC MODEL

Having defined in general terms the concepts of simulation experimentation, we can now turn to the detailed conduct of individual experiments. The amount of information we need to obtain from any experiment affects the experimental design that we adopt for that experiment. The simplest

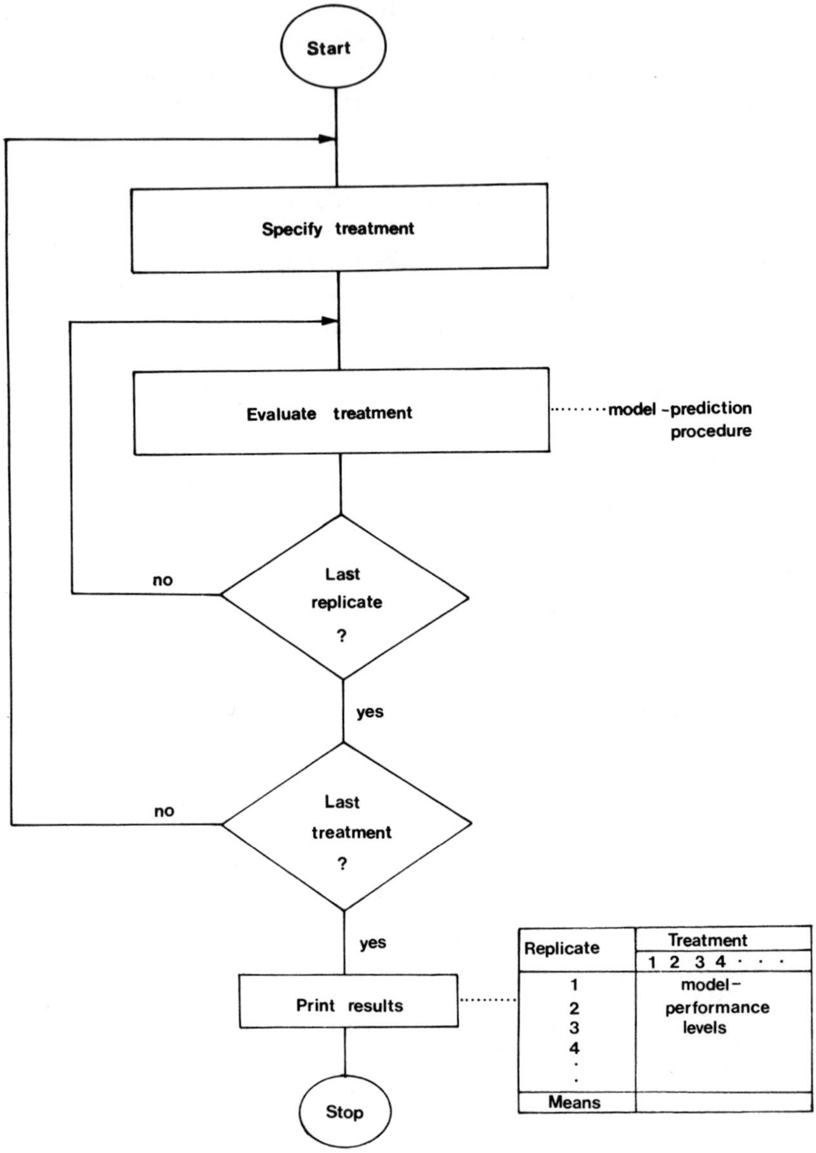

Fig. 6.1. Procedure for experimentation with a computer model of a bioeconomic system. Notes. 1. Each replicate of each treatment requires one full run (or encounter with) the model; 2. The entire experiment is carried out within a single computer run.

case is where performance Z depends on a single controllable factor x, and no non-controllable factors are included in the model, i.e.,

$$Z = f(x)$$

and x takes only two values.

For example, in a disease-control experiment using the barley-leaf-rust model we may wish to consider the two alternatives of spraying versus no spraying. Here we may let

$x = 0$ represent 'no spray';

$x = 1$ represent 'spray'.

There is no need to replicate these treatments since identical results would be obtained for each replicate when using a deterministic model. Each treatment is evaluated once, performance being predicted in terms of, say, total cost of labour, purchased chemicals and crop-loss due to disease. Let us call these criterion values Z_1 and Z_2 for 'no spray' and 'spray' respectively. No statistical analysis of the results is needed; the decision rule is simply

if $Z_1 < Z_2$ then treatment $x = 0$ is superior and no spraying is recommended;

if $Z_1 > Z_2$ then treatment $x = 1$ is superior and spraying is recommended.

However, while such an experiment is simple to conduct and interpret, the information provided is extremely limited. The assumption here is that spraying is uniformly effective over the whole range of conditions. While this may be true in some circumstances, more frequently we would expect exogenous variables such as weather to affect the outcome Z to some significant extent. Therefore we need to consider experimentation with models in which these stochastic or uncertain variables are explicitly included.

SINGLE-FACTOR EXPERIMENTS—STOCHASTIC MODEL

If the model is stochastic, then levels of some non-controllable factors are obtained by sampling from their respective probability distributions or from historical time-series data. The performance criterion now depends not only on the treatment imposed but also on the particular environmental sequence involved. Disease build-up may depend, for example, on dew conditions during the growing season of the crop. The simulation experimenter has some control over the environment in which the model operates—a control which is not within the reach of the conventional experimenter. When historical time-series data are being used directly to establish the environment, treatments are simply compared under identical segments of the whole series. Replicates are formed by using different segments of the time series under which to explore each treatment. Alternatively, control can be exerted over a generated sequence. Recall from Chapter 4 that each random number from a uniform distribution is obtained from the previous random number, and that once the seed is chosen the sequence of numbers is fully determined. Further, a given sequence of uniformly distributed random numbers results in a given sequence of values from the target distribution. Where environmental data are obtained by sampling from probability distributions, it is important to specify the manner in which the pseudorandom-number generator is seeded. Two options are available:

1. Identical seeding. When the seed (NSEED in Chapter 4) is reset at the same value for each new treatment, the same environmental sequence will be generated for corresponding replicates under each treatment. For example, if a drier-than-average season is generated for the first replicate under no spraying, then this same drier-than-average season will occur under the first replicate of the spraying alternative, and so on.
2. Independent seeding. By resetting the seed in a random fashion before sampling for each stochastic variable for each treatment, different environmental sequences will be generated for the different treatments within the same replicate.

In the case of the barley-leaf-rust model, the relationship to be investigated may now be written as

$$Z = \mathrm{f}(x, y)$$

where y is the dew variable. (More precisely, dew in each time period (day in the case of the barley-leaf-rust model) throughout the crop growing season is a separate stochastic variable (y_1, y_2, $y_3, \ldots$) to be sampled from a separate probability distribution.)

To compare the treatments it is now necessary to repeat (replicate) each a number of times under different simulated dew sequences, and to take as our criterion the mean performance over all replicates, i.e.

$$Z_1 = \frac{\text{sum of cost levels over all replicates of no spraying}}{\text{number of replicates of the 'no spray' treatment}}$$

and similarly for Z_2. These two means may now be compared using the

Student t-test. The null hypothesis of this test is that the two means are equal; the alternative hypothesis may be that

1. they are unequal;† or
2. one is greater than the other; or
3. one is the greater by a minimum amount of economic significance, \$$A$.

Under certain circumstances an alternative test requiring fewer replicates to achieve significance is available to determine which treatment is best. This is the t-test on differences in performance between paired replicates. Which of the tests is appropriate depends on the method employed to obtain values for the non-controllable variables (here dew) and in particular on the manner of seeding the pseudorandom-number generator. Under independent seeding, differences in mean performance Z may be due to treatment effects or replicate (i.e. environment) effects. Identical seeding, by allowing us to evaluate the treatments under identical environments, eliminates this latter source of variability and sharpens the comparison between treatments. When identical seeding is adopted, corresponding replicates may be regarded as paired observations and we may test whether the mean difference between performance under these paired observations is significantly different from zero,‡ or from some fixed amount \$$A$.

EXPERIMENTS WITH A SINGLE FACTOR AT MANY LEVELS

To extend our experiment a stage further, suppose the controllable factor can take not just two but several levels, e.g. several fungicide-application

† The hypothesis H_0: $\mu_1 = \mu_2$ is compared against the alternative of unequal treatment effects on the basis of the test statistic

$$t = \frac{Z_1 - Z_2}{\sqrt{\dfrac{(n_1 - 1)s_1^2 + (n_2 - 1)s_2^2}{n_1 + n_2 - 2}}\sqrt{\dfrac{1}{n_1} + \dfrac{1}{n_2}}}$$

where μ_1 and μ_2 are the expected performances under each treatment, n_1 and n_2 are the numbers of replicates (usually equal) and s_1^2 and s_2^2 are the variance estimates. The test statistic has $n_1 + n_2 - 2$ degrees of freedom.

‡ The hypothesis H_0: $\mu_0 = 0$ is tested on the basis of the statistic

$$t = \frac{\bar{D}}{s_{\bar{D}}/\sqrt{n}}$$

where μ_0 is the expected difference between treatment effects, $\bar{D}$ is the observed mean difference in performance between paired replicates, $s_{\bar{D}}$ is the standard deviation of differences, and n is the number of replicates. The test statistic has $n - 1$ degrees of freedom.

rates. Each treatment can now be defined symbolically as a specified quantity of fungicide applied per hectare.

$$Z = f(x_i, y)$$

where x_i, $i = 1, \ldots n$ are the individual fungicide application rates (treatments) chosen for the experiment and n is the number of treatments. All treatments are evaluated and the value for x which gives the greatest Z is chosen as the optimal value. Alternatively, and to increase precision, a least-squares regression may be used to fit a response curve relating Z to fungicide-application rate x.

This curve could be quadratic of the form

$$Z = m_0 + m_1 x + m_2 x^2$$

The mathematical rules for locating extreme values of functions now tell us that this cost curve is stationary (horizontal) at

$$x = -\frac{m_1}{2m_2}$$

and that this is the point of optimality provided $m_2 > 0$.

Once again, if the model is stochastic, each treatment must be replicated a number of times. In this case, the model-output may be subjected to either regression analysis or a one-way analysis of variance (ANOVA). The former procedure, applied to either individual replicate performance or treatment means, again allows us to interpolate between treatments to the optimal fungicide level. ANOVA, on the other hand, is used simply to compare the discrete treatments of the experiment, indicating which ones have mean costs significantly lower than the others. Here again, the method of seeding influences the interpretation of model-results. When identical seeding is adopted the replicates under different treatments do not form independent samples. Since ANOVA assumes independent samples, its use is not applicable under identical seeding (Phillips, 1971).

TWO EXPERIMENTAL FACTORS—DETERMINISTIC MODEL

In many instances of interest to us, two or more factors or variables will be under management's control. To experiment with such a system is somewhat more complex and in the following sections we will be discussing two-variable experiments and then multifactor experiments.

> Suppose a property contains 150 ha of arable land and the farmer wishes to know what areas of wheat and potatoes to grow in order to obtain the highest

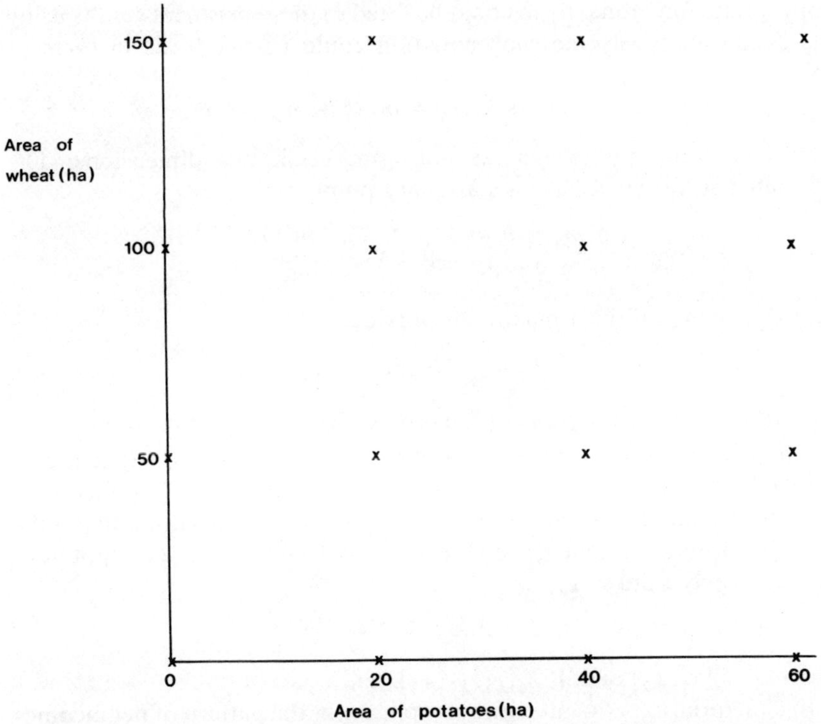

Fig. 6.2. *An example of a factorial design with two factors each at four levels.*

possible net income. (Land not used for crops may be placed under pasture.) The model now becomes

$$Z = \mathrm{f}(X) \text{ and } X = [x_1 x_2]$$

where Z is farm net income, x_1 is the area of wheat, and x_2 is the area of potatoes.

A simulation experiment with a model of this farm system could be carried out using a design such as in Fig. 6.2. Here each cross represents a design point or treatment. Each of four levels of x_1 is combined with each of four levels of x_2 giving rise to 16 treatments spaced evenly over the 'policy space' or spectrum of feasible management alternatives. While wheat areas range up to the full 150 ha, labour and machinery availability restrict the area of potatoes to 60 ha.

The experiment is conducted along the lines of Fig. 6.1, but at this stage without replication. Having evaluated all the treatments we may select the one for which predicted income is greatest as the optimal policy. Alternatively, as with the single-factor experiment, an equation with an appropriate functional form could be fitted to the experiment results using regression analysis. One such equation could be quadratic in form as follows:

$$Z = m_0 + m_1x_1 + m_2x_2 + m_3x_1^2 + m_4x_2^2 + m_5x_1x_2$$

This equation represents a response surface in two dimensions. The estimated surface will have a stationary point at

$$x_1 = \frac{2m_1m_4 - m_2m_5}{m_5^2 - 4m_3m_4} \text{ and } x_2 = \frac{2m_2m_3 - m_1m_5}{m_5^2 - 4m_3m_4}$$

and this point will be a maximum provided

$$m_3 < 0,\ m_4 < 0 \text{ and } m_3m_4 > m_5^2$$

TWO EXPERIMENTAL FACTORS—STOCHASTIC MODEL

In Chapter 4 we discussed the effects of risk and uncertainty on agricultural decision making and we can extend the discussion to include these factors in an experimental design. Suppose that net income depends not only on the cropping programme but also on the price of wheat y_1, the price of potatoes y_2 and weekly rainfall y_3, i.e.

$$Z = f(X, Y)$$

where $X = [x_1, x_2]$ and $Y = [y_1, y_2, y_3]$.

In a factorial experiment without replication, the pattern of net incomes due to the various treatments imposed would be obscured by random variations due to the different bioeconomic environments. We could, of course, evaluate each cropping policy under the same set of exogenous variables of price and rainfall regimes. This could be achieved by identical seeding of the appropriate random-number generators. Obviously this has the disadvantage that our findings could not be generalised to other circumstances. A better approach is to replicate the treatments—i.e. to evaluate each combination of crop areas under a number of different sets of exogenous variables. The performance criterion for a given treatment is now based on the outcome over all replicates (e.g. mean net income, or in a Bernoullion decision theory framework, a weighted function of mean income and variance in income). The results of the experiment may again be analysed by a surface-fitting approach, or by analysis of variance. The latter considers only discrete policies (treatments) rather than interpolating between them, but allows an examination of significant differences between policies. The requirement for ANOVA of independent seeding should again be noted.

EXPERIMENTS WITH MORE THAN TWO FACTORS

The above procedure may be generalised for multifactor simulation experiments. Consider an example with two factors, each at four levels, i.e. a 4×4 or 4^2 full factorial design. Replicating each treatment, say, 10 times, gives rise to 160 performance predictions or encounters with the model. While this full factorial design is suitable where we have a relatively simple experiment, it becomes unwieldy as the number of factors and/or levels increases. Suppose now that there are seven policy variables, and each is to be evaluated at four levels, a minimal number for acceptable response surface fitting. The full factorial design would then include 4^7 or 16,384 design points or treatments, each of which would need to be replicated, say, 10 times. While the cost of this experiment would vary with the size and complexity of the computer model and pricing policy of the particular data-processing installation, let us make some not unreasonable minimal assumptions in order to arrive at a cost estimate. Supposing each encounter with the model requires 1 s of processor time at a charge of \$100/h; the cost of the experiment would be approximately \$4,600.

Cost can be reduced by using a partial or incomplete factorial design, omitting some combinations of factor levels and assuming that some of the interaction effects are zero. A partial factorial may require as few as one-sixteenth of the number of treatments of the full factorial. Rotatable and response surface designs allow even greater reduction in the number of treatments required. These traditional designs have an important role to play in experimentation with systems models, particularly where the number of controlled factors is not more than four. They are also useful for exploratory experimentation or preliminary screening of factors to determine which ones have the greatest effect on performance. Since traditional designs are adequately explained in a number of textbooks (e.g. Box, 1966; Cochrane & Cox, 1957; Mendenhall, 1968), it is not necessary to cover the same ground here. Instead, we will consider a special group of less-well-known designs which allow an even greater reduction in treatment numbers and hence allow optimisation with respect to as many as 10 or 15 policy variables.

Optimum-seeking Designs

Optimum-seeking designs have their origin in mathematical techniques of numerical optimisation which have been devised for such purposes as solving complex systems of non-linear equations. These methods are not usually applicable to the design of real-life experiments with bioeconomic systems, though often useful in engineering applications, and depend on the special features of simulation experiments referred to earlier. In each of the designs which have been discussed in the preceding sections the complete layout of the experiment is specified before the experiment commences. This is traditionally the case in real-life biological experiments where the treatments are evaluated simultaneously, e.g. different fertiliser rates are applied to plots of plants where each plot is sown on the same day and each is harvested on the same day: the entire experiment is conducted in one season or year. In the case of simulation experiments, each encounter with the model takes a very short period of time, we are able to reproduce an environment exactly, and of course the mode of operation of the computer is to evaluate treatments one at a time, that is, sequentially rather than simultaneously. In a simulation experiment it is, therefore, practical to use information gained from earlier treatments in guiding the placement of later treatments and thus to exclude unpromising regions of factor levels from further consideration. A set of rules for placement of treatments given the information obtained to date can be written into the computer model, most likely as a subroutine coupled to the main program of the systems model. Identical seeding, which is precluded under traditional designs relying on the analysis of variance technique, can be used to advantage to sharpen the contrast between treatments and thereby reduce the number of replicates required. The elimination of wasted treatments and reduction of the number of replicates means that a substantial increase in the number of experimental factors or, conversely, a marked reduction in the number of treatments is possible. As an approximate guide, our seven-factor experiment mentioned previously requiring 16,384 treatments in factorial design may be replaced by as few as 160 treatments each replicated approximately five times. Optimum-seeking designs therefore have an important role to play in effective and efficient simulation experimentation. We will be discussing a number of these designs in general terms in this chapter. For a detailed discussion of the mathematics and the theory of optimum-seeking designs, as well as a set of subroutines for their execution, the reader is referred to Harrison (1978).

A good optimum-seeking method should possess the qualities of robustness, efficiency and precision. A robust procedure is capable of locating optimal or near-optimal values for a wide range of problems, regardless of the shape of the response hypersurface. This statement must be qualified by noting that the surface should be well-behaved in the sense of having a unique optimum. If it is multimodal (cf. a mountain range with a number of separate peaks) then convergence to a local rather than global optimum may take place. This problem can only be overcome by carrying out a number of optimum-seeking experiments with different starting treatments. A design method is efficient if it is able to locate near-optimal values with a small number of treatments. Precision refers to the accuracy with which the optimum indicated approximates the true but unknown system optimum.

THE METHOD OF STEEPEST ASCENT

The problem of locating the maximum point on a response surface relating performance of a system to levels of two or more controllable factors may be likened to the physical analogy of a person climbing to the top of a mountain, and in fact the techniques employed are often referred to as 'hill-climbing' methods. One of the oldest and simplest of these design procedures is the method of steepest ascent. In essence this method establishes the shape of the response surface at a given point, determines the direction of steepest incline, and places successive treatments at uniform intervals in this ascent direction. Once a crest has been reached, a new steepest-ascent direction is determined and a new series of steps taken, the whole procedure being repeated until no further progress is possible.

The experiment commences by the experimenter choosing an initial treatment or policy as close as possible to the optimum, based on prior knowledge about the system. For simplicity, we may take the case of two experimental factors or controllable variables, and designate this first treatment as

$$X^0 = [x_1^0, x_2^0]$$

where x_1^0 and x_2^0 are the initial guesses of the optimal values of the variables x_1 and x_2. The simulation model is used to evaluate this treatment and predicts a performance level of Z^0.

In the earlier example, suppose the initial guess is to grow 60 ha of wheat and 20 ha of potatoes and that net income for this policy, averaged over all replicates, is $38,000. The policy space, response surface (the shape of which is unknown to the researcher) and initial treatment may be illustrated as in Fig. 6.3. In part (a) of this figure response is depicted by a surface diagram in three dimensions, Z, x_1 and x_2. The performance or Z-axis has been eliminated in part (b) to leave an x_1–x_2 plane, similar to Fig. 6.2. However, response contours or lines joining points of equal farm net income have been drawn on this diagram. The line of steepest ascent from any contour is perpendicular to the contour at that point.

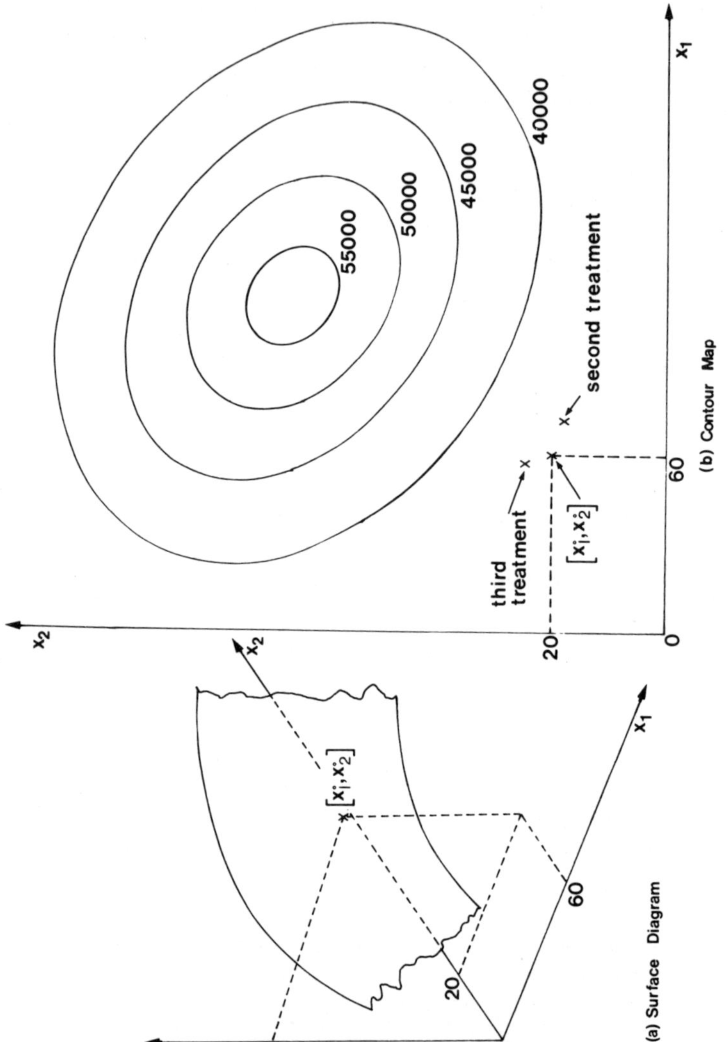

Fig. 6.3. Diagrammatic representation of policy space and response surface.

Next (but within the same computer run) two further treatments are evaluated. These treatments are the result of small increments to x_1 and x_2 in turn:

$$X^1 = [x_1^0 + \Delta x_1, x_2^0]$$

$$X^2 = [x_1^0, x_2^0 + \Delta x_2]$$

The performance values Z^1 and Z^2 resulting from these two treatments are used to estimate the slopes of the response surface (m_1 and m_2) with respect to x_1 and x_2, i.e.

$$m_1 = \frac{Z^1 - Z^0}{\Delta x_1} \quad \text{and} \quad m_2 = \frac{Z^2 - Z^0}{\Delta x_2}$$

This information makes it possible to determine the direction of steepest ascent. Stepping using a defined step size is continued in this direction as long as performance Z continues to improve as shown in Fig. 6.4. The first step for which Z declines (treatment 8 in Fig. 6.4) is discarded and the previous or

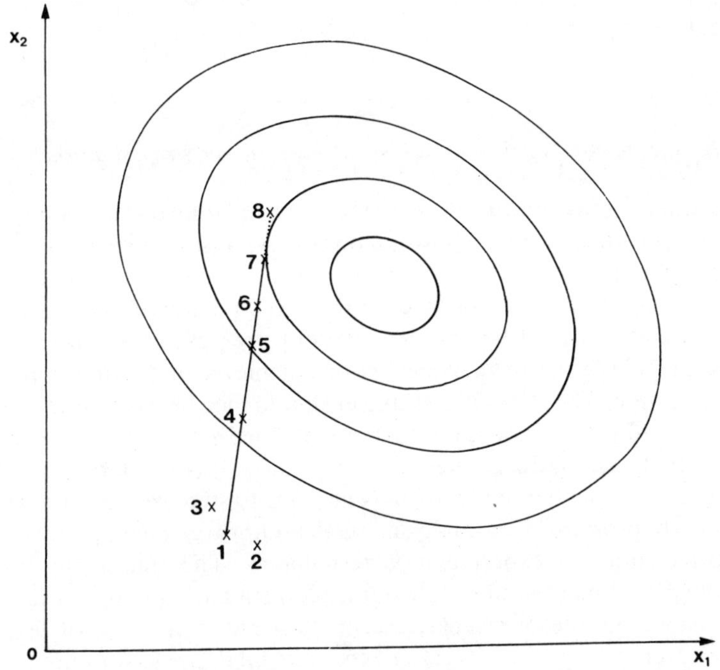

Fig. 6.4. Placement of treatments in first iteration of steepest ascent.

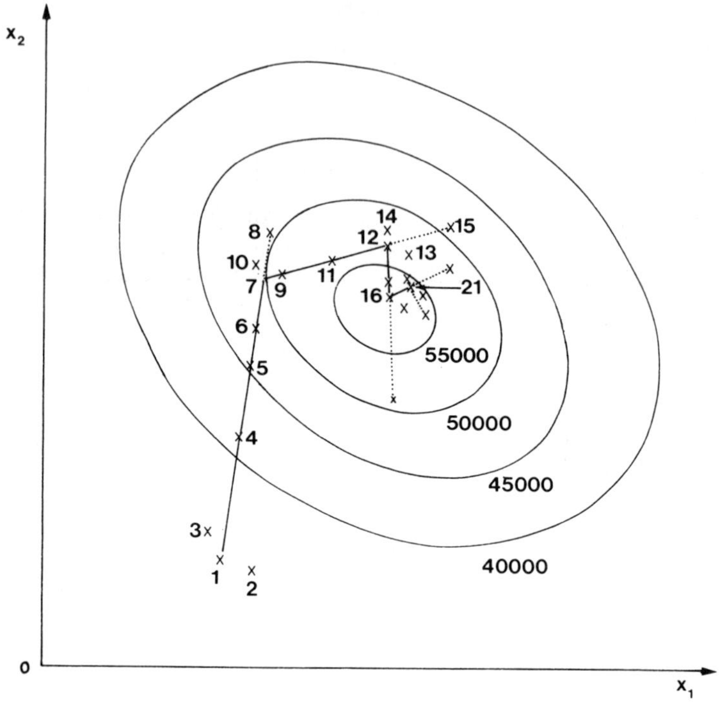

Fig. 6.5. Complete experimental design under steepest ascent.

best treatment is taken as a new search base, completing one iteration of the procedure. At this improved position on the response surface the variables should again be incremented in turn, the direction of steepest ascent estimated and a series of steps taken in the new direction of steepest ascent. Further iterations may be carried out until progress ceases because no improvement in the performance criterion is achieved on the first step of the ascent direction. The step-size parameter should then be reduced and more-closely-spaced treatments in the currently defined search direction should be evaluated. The reduced step size may allow further iterations to be carried out, and further reduction may be made to step size when the search again fails to progress. Eventually, no further improvement in performance is possible and the experiment is terminated. The treatments of the complete experiment would trace out a path similar to that of Fig. 6.5.

The above outline of steepest ascent illustrates a number of features common to most optimum-seeking experimental designs. An initial guess of the optimal policy is made and the search proceeds through a number of iterations. Each iteration uses a fixed set of rules for placement of treatments, and iterations cease when no further progress can be made (or a stopping rule based on precision or cost is applied). The speed with which the optimum is approached depends on the shape of the response surface, closeness of the initial guess and settings of the search parameters (e.g. step size). Appropriate search-parameter settings can only be determined by judgement and by experience from carrying out exploratory experiments with the model.

Steepest ascent, while a relatively easy method to understand and use, is not as robust, efficient or precise as some other hill-climbing methods. One particular deficiency results from the fixed step size which may result in hopping backwards and forwards across a ridge when variables are interacting (and response contours are flattened ellipses).

A number of procedures other than steepest ascent have been developed for numerical optimisation, and only a few offering the greatest promise for bioeconomic-systems studies will be mentioned here. Further details of these procedures can be found in Harrison (1978).

1. *The simplex method*

This method, not to be confused with a solution algorithm for linear programming of the same name, operates through use of a 'moving simplex'. A simplex in an n-dimensional space is a figure having plane sides and $n + 1$ vertices, e.g. a triangle in the x_1–x_2 plane. Use of the simplex method for optimisation of economic systems has been investigated by Meier (1967) and by Meier, Newell and Pazar (1969).

The concepts behind this method may be explained graphically and without algebra by reference to Fig. 6.6. Here the first three treatments (a, b and c) are spaced equal distances apart at the three vertices of a triangle. This triangle can be moved uphill by 'reflection' or flipping over in a direction opposite the lowest vertex. Thus, if treatment evaluations with the systems model reveal that Z_a is less than both Z_b and Z_c then treatment (a) will be discarded and a new triangle formed by placing treatment (d) equidistant from (b) and (c) on the opposite side of the (b–c) face. Suppose now that Z_b is less than Z_c and Z_d. A new treatment is placed at (e). The triangle is moved up the response surface in this manner until treatment (k) in the vicinity of the summit has been evaluated. At this stage, further search involves revolution of the triangle about the region of optimality, and the search is concluded after one such revolution.

Modifications to the basic procedure such as changing the size or shape of the simplex during the search have been found to increase efficiency and

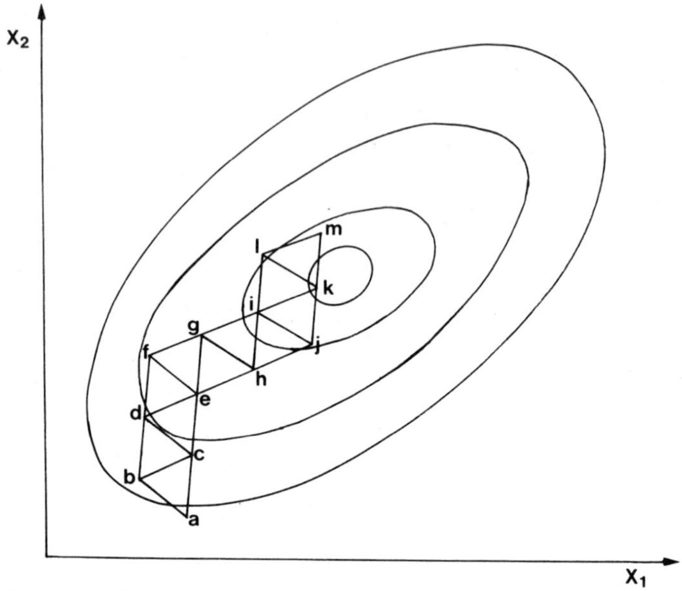

Fig. 6.6. Steps in bivariate optimisation using the simplex method.

precision. However, evidence suggests that the simplex method is not very satisfactory when there are more than three policy variables (Box, 1966).

2. *Conjugate directions*

The method of conjugate directions has been devised by Powell (1964) to locate the minimum of a general function of n variables and used in systems studies by Harrison & Longworth (1977). This orientation towards minimisation presents no problem when we are seeking maxima; values of the controllable variables maximising performance are obtained if we simply change the sign of the criterion and minimise $-Z$. Conjugate-direction search has the important property of quadratic convergence, that is, the exact optimum of a quadratic form will be found in a finite and known number of iterations. Since the response surfaces of many bioeconomic systems can be closely approximated by quadratic functions in the regions of their optima, the method is quite robust. As well, comparative studies indicate that it ranks highly in terms of efficiency (Powell, 1964).

3. *Random search*

This method consists basically of specifying a range for each variable and ascribing a uniform distribution over this range. Each treatment is obtained by using random numbers to select values of the factors from their respective ranges. A large number of treatments are evaluated and a sorting procedure is applied to the results to determine the treatment with the highest criterion value.

The precision of random search tends to be low unless the number of treatments is large and the method has been found to be relatively inefficient, even by comparison with fixed designs such as factorials.

4. *Random search with learning*

Substantial improvement can be made to efficiency of random search through 'heuristic learning' or revision of probabilities during the experiment. This involves progressively reshaping the (initially uniform) probability distributions by increasing the probabilities of those factor levels for which high performance is predicted. Random search with learning has been used with apparent success in a number of farm-planning studies (Boehlje, 1973; Eisgruber & Lee, 1971; Furtan & Lee, 1975), important advantages over non-random methods including

1. it is conceptually simple, not relying on difficult mathematics, and is easy to programme for a computer;
2. controllable variables which are confined to integer levels can be handled without difficulty;
3. variables may be made mutually exclusive, complementary or conditionally complementary;
4. no matter how many variables are included, a solution is obtained (whereas other methods may make no progress) and this solution is usually at least reasonable; and
5. poorly shaped (including multimodal) responses surfaces are handled more successfully than under alternative design procedures.

To explain the method, suppose we have two controllable variables x_1 and x_2 and each can take a number of mutually exclusive discrete values. Each treatment in the experiment is obtained by sampling from the cumulative probability distributions of the variables. The search proceeds in three stages. The first or initialisation phase involves repeated sampling from uniform probability distributions which are defined for each variable and the value of the performance criterion is evaluated for each set of values of the variables. The highest criterion value in the initialisation phase is taken as a base performance (Z_0) against which to compare alternative policies in the second or learning phase.

In the learning phase, the probabilities of selecting different values of the policy are adjusted according to a formula including the performance criterion Z and a rate-of-learning parameter. This last parameter is specified by the experimenter and must be chosen to suit the particular problem. If the adjustment of probabilities is too slow, then the search will be inefficient and unduly costly. On the other hand, rapid adjustment may lock the search into a local optimum rather than seeking out a global optimum. The learning phase is continued for a fixed number of treatments, producing distributions for each variable which may be quite peaked and skewed. In the final sampling stage the probability distributions are locked in and a set number of treatments are evaluated. The treatment resulting in the highest criterion value in this final sampling phase is selected as the optimal policy. The number of treatments in each of the three stages also has an important bearing on efficiency of the search. These numbers, and the learning-rate parameter, can only be decided after a number of trial searches with the model.

Concluding Comments

The foregoing discussion provides a brief introduction to experimental design and analysis procedures for implementation of bioeconomic simulation models. These procedures are summarised in Table 6.1 which is intended as an approximate guide only. In general, the recommended designs range from complete enumeration of alternatives, through partial enumerative designs specified prior to experimentation and finally to optimum-seeking designs. Appropriate fixed designs can be obtained from books on experimentation in agriculture (Cochran & Cox, 1957; Heady & Dillon, 1961; Mendenhall, 1968). When using an optimum-seeking design it is necessary to either write a computer routine for the procedure or to call upon a program already developed for this purpose (Harrison, 1978; Hopper, 1973; Keuster & Mize, 1973).

Constructive use of experimental designs, and of statistical analysis of results where applicable, provides for the simulation modeller a facility to determine optimal management policies or optimum combinations of factors. The model we have put so much effort into developing can then be used with maximum efficiency.

Table 6.1. Suggested design and analysis procedures

Type of model	*Stage of implementation*	*Number of experimental factors or controllable variables*				
		1		2	3 or 4[a]	4 or more[a]
		2 levels	More than 2 levels			
Deterministic model	Design	Evaluate each treatment		Complete factorial		Optimum-seeking design[c]
	Analysis	Select treatment giving best performance or regression analysis			Regression analysis	None
Stochastic model	Design	Evaluate all replicates of all treatments			Fixed design (incomplete)[b]	Optimum-seeking design[c]
	Analysis	*t*-test	One-way ANOVA or regression analysis	Two-way ANOVA or regression analysis	General ANOVA or regression analysis	None

[a] Either fixed or optimum-seeking designs may be found preferable in the overlapping case of four factors.
[b] For example, partial factorial or central composite design.
[c] For example, conjugate directions or random search with learning.

Appendix

This appendix sets out the computer routine for the experimental procedure method of steepest ascent. Other computer routines for experimental approaches mentioned in Chapter 6 can be found in Harrison (1978). The program presented is written in FORTRAN for a Burroughs B6700 computer but is designed for ease of adaptation to other makes of machine. The operating system under which the programs were developed has unusual features with respect to line-spacing control and loss of constants (but not arrays) in a subroutine between successive calls, and these are overcome by adding dummy WRITE statements and COMMON statements respectively.

The FORTRAN subroutine for the procedure follows the layout presented in Fig. A6.1. The subroutine initially reserves space for arrays (in a DIMENSION statement), specifies search parameters (as DATA), makes type declarations (INTEGER and REAL) and indicates which constants and arrays are common to the main program and subroutine. The first time the subroutine is called, a number of initial conditions are established. At the heart of the layout in Fig. A6.1 is a branching GO TO statement conferring control to different segments of the subroutine depending on the value of a test criterion, *ITEST*. Each segment carries out a specific part of the search and ends with a RETURN to the main program where the treatments are evaluated. Branching to a given segment is repeated for a number of treatments, until that particular phase of the search iteration is completed, e.g. control returns to the segment placing treatments in the direction of steepest ascent until the objective function falls relative to the previous treatment.

None of the programs contain READ statements, all information for the search being defined in the DATA statements.

The subroutine is designed for coupling to a systems model-program with little or no reprogramming. However, the values of the initial policy and search parameters as defined in DATA statements would need to be adjusted on a trial-and-error basis until settings providing the most efficient design for the particular application were ascertained.

Output of results is formatted with row rather than column headings to avoid confusion with any other results from the experiments which the user may wish to have printed. The test function (surrogates for the systems model) in the example is a simple polynomial predicting dependent variable Z and with a known optimum.

```
   Type statements (REAL and INTEGER), DIMENSION, COMMON,
   DATA.
   IF (ITREAT.GT.1) GO TO 2
   .
   .
   . Initial conditions
   .
   .
 2 GO TO (30, 40, 50), ITEST
   .
   .
   . Establish slopes of the targent hyperplane
   .
   .
   ITEST = 1
   RETURN
30 . . . . .
   .
   .
   . Step in the direction of steepest ascent
   .
   .
   ITEST = 2
   RETURN
40 . . . . .
   .
   .
   . Take a step backwards
   .
   .
   ITEST = 3
   .
   .
   . Reduce step size
   .
   .
   ITEST = 0
   RETURN
50 . . . . .
   .
   .
   . Terminate search
   .
   .
   RETURN
   END
```

Fig. A6.1. Layout of the steepest-ascent subroutine.

The parameters as listed in the DATA statements have the following interpretations:

Main program

$$X = \text{initial treatment, here} \begin{bmatrix} 5{\cdot}0 \\ -8{\cdot}0 \\ 10{\cdot}0 \end{bmatrix}$$

$TMAX$ = maximum number of treatments before the search must terminate (here 40).

Subroutine

$NVAR$ = number of controllable variables (3)
$S(1), S(2), S(3)$ = forwarded difference for variables 1 to 3, here all 0·1.
L = step-size parameter (2·0)
$NRED$ = number of reductions in step size (1)
$SRED$ = extent of reduction in step size (5·0)

Main Program

```
FILE  5=FILE5,UNIT=READER
FILE  6=FILE6,UNIT=PRINTER

      INTEGER TMAX
      DIMENSION X(10)
      COMMON X
     1,IVAR,ITEST,ISTEP,ZL,DEN,ITN
      DATA IN/5/,10/6/
      DATA X(1)/5./,X(2)-8./,X(3)/10./,TMAX/40/
      ITREAT = 0
      ITREAT = ITREAT + 1
      Z = (X(1)-2.*X(2))**2 + (X(2)-3.*X(3))**2 +
          (3.*X(1)-2.*X(3))**2
     1- 2.*X(1) + X(2) - 3.*X(3) + 10.
      Z = -Z
      CALL OPT2(ITREAT,TMAX,Z)
      IF (ITREAT.LT.TMAX) GO TO 1
      STOP
      END
```

Subroutine

```
C  SUBROUTINE TO MAXIMIZE A FUNCTION OF SEVERAL
C  VARIABLES USING THE METHOD OF STEEPEST ASCENT

      SUBROUTINE OPT2(ITREAT,TMAX,Z)
      INTEGER TMAX
      REAL M,L
      DIMENSION X(10),S(10),M(10),D(10)
      COMMON X
     1,IVAR,ITEST,ISTEP,ZL,DEN,ITN
      DATA NVAR/3/,S(1)/.1/,S(2)/.1/,S(3)/.1/,L/2./,NRED/1/,SRED
     /5./,1 10/6/
      IF (ITREAT.EQ.1) WRITE (10,4)
4     FORMAT (1X,'INITIAL ESTIMATE'/)
      WRITE (10,6) ITREAT,Z,(X(J),J=1,NVAR)
6     FORMAT (3X,'TREATMENT NO',13,4X,'Z =',F10.4,4X,
     'XVALUES:',6F10.4/1 3X,4F10.4/)
      IF (ITREAT.GT.1) GO TO 2
      ITEST = 0
      IVAR = 0
      ITN = 1
      WRITE (10,9) ITN
9     FORMAT (/1X,'ITERATION NO',12/)
      ZL = Z
2     GO TO (30,40,50), ITEST
      IF (IVAR.EQ.0) GO TO 22
C  DERIVE EQUATION TO TANGENT HYPERPLANE
      M(IVAR) = (Z-ZL) / S(IVAR)
      X(IVAR) = X(IVAR) - S(IVAR)
      WRITE (10,13) Z,ZL,M(IVAR)
13    FORMAT (3X,'CURRENT Z VALUE =',F15.4, 8X,'BASE
     ZVALUE =',F15.4,1 8X,'SLOPE =',F15.4)
22    IVAR = IVAR + 1
      X(IVAR) = X(IVAR) + S(IVAR)
      IF (IVAR.EQ.NVAR) ITEST = 1
      WRITE (10,12) IVAR,S(IVAR),X(IVAR)
12    FORMAT (3X,'VARIABLE',13,' FORWARD DIFFERENCED
     BY ', F9.4,' TO',1 F10.4)
      RETURN
```

```
C
C  DETERMINE DIRECTION OF STEEPEST ASCENT
30     M(NVAR) = (Z-ZL) / S(NVAR)
       WRITE (10,13) Z,ZL,M(IVAR)
       X(NVAR) = X(NVAR) - S(NVAR)
       SSQ = M(1)*M(1)
       DO 32 IVAR=2,NVAR
32     SSQ = SSQ + M(IVAR)*M(IVAR)
       DEN = SQRT(SSQ)
38     ISTEP = 0
C  STEP IN STEEPEST ASCENT DIRECTION
       DO 34 IVAR=1,NVAR
34     D(IVAR) = M(IVAR) / DEN * L
       WRITE (10,14) (D(J),J=1,NVAR)
14     FORMAT (3X,'ADJUSTMENTS TO X VALUES:',8F9.4)
36     ISTEP = ISTEP + 1
       WRITE (10,15) ISTEP
15     FORMAT (3X,'STEP NO',I3)
       DO 35 IVAR=1,NVAR
35     X(IVAR) = X(IVAR) + D(IVAR)
       ZL = Z
       ITEST = 2
       RETURN
C
40     CONTINUE
       IF (Z.GT.ZL) GO TO 36
       ISTEP = ISTEP - 1
       WRITE (10,16) ISTEP
16     FORMAT (3X,'X VALUES AT STEP',I2,1X,'TAKEN AS
     NEW SEARCH BASE')
C  FUNCTION VALUE DECREASING. STEP BACKWARDS
       Z = ZL
       DO 42 IVAR=1,NVAR
42     X(IVAR) = X(IVAR) - D(IVAR)
       IF (ISTEP.GT.1) GO TO 44
C  FUNCTION HAS DECREASED ON FIRST STEP
       IF (NRED.GT.0) GO TO 46
       ITREAT = TMAX - 1
       ITEST = 3
       RETURN
```

```
C
C  REDUCE STEP SIZE
46     L = L / SRED
       NRED = NRED - 1
       WRITE (10,19) L
19     FORMAT (3X,'STEP SIZE PARAMETER REDUCED TO',
      F10.4)
       GO TO 38
44     IVAR = 1
       ITN = ITN + 1
       WRITE (10,9) ITN
       X(IVAR) = X(IVAR) + S(IVAR)
       WRITE (10,12) IVAR,S(IVAR),X(IVAR)
       ITEST = 0
       RETURN
C
50     ITREAT = TMAX
       RETURN
       END
```

References

Boehlje, M. (1973). Optimisation and decision models: the use of statistical search procedures. *Can. J. Agr. Econ.*, **21,** 43–53.

Box, M. J. (1966). A comparison of several current optimisation methods, and use of transformations in constrained problems. *Computer J.*, **9,** 66–77.

Cochran, W. G. & Cox, G. M. (1957). *Experimental Designs*, 2nd ed. New York: Wiley.

Eisgruber, L. M. & Lee, G. E. (1971). A systems approach to studying the growth of the farm firm. *In:* Dent, J. B. & Anderson, J. R. (eds.), *Systems Analysis in Agricultural Management*. Sydney: Wiley.

Furtan, W. A. & Lee, G. E. (1975). An example of a heuristic learner in a simulation model. *Can. J. Agr. Econ.*, **23,** 31–40.

Harrison, S. R. (1978). *Optimum-Seeking Designs for Simulation Experiments with Models of Agricultural Systems.* Agricultural Economics Research Unit Research Report No. 87.

Harrison, S. R. & Longworth, J. W. (1977). Optimal growth strategies for pastoral farms in the Queensland Brigalow scheme. *Australian J. Agr. Econ.*, **21,** 80–96.

Heady, E. O. & Dillon, J. L. (1961). *Agricultural Production Functions.* Ames: Iowa State University Press.

Hopper, M. J. (ed.) (1973). *Harwell Subroutine Library: A Catalogue of Subroutines.* London: HMSO.

Meier, R. C. (1967). The application of optimum-seeking techniques to simulation studies: a preliminary evaluation. *Financial Quant. Anal.*, **2,** 11–51.

Meier, R. C., Newell, W. T. & Pazar, H. L. (1969). *Simulation in Business and Economics.* Englewood Cliffs: Prentice-Hall Inc.

Mendenhall, W. (1968). *Introduction to Linear Models and the Design and Analysis of Experiments.* Belmont: Wadsworth.

Phillips, J. B. (1971). Statistical Methods in Systems Analysis. *In:* Dent, J. B. & Anderson, J. R. (eds.), *Systems Analysis in Agricultural Management.* Sydney: Wiley.

Powell, M. J. D. (1964). An efficient method for finding the minimum of a function of several variables without calculating derivatives. *Computer J.* **7,** 155–62.

7

Model-Application

Increasingly the scientific community is being urged towards multidisciplinary action on a project basis and the forces fuelling this drive are similar to those embedded in the systems concept. The basic hypothesis is that a system—whether defined as a physiological system for a plant or the economic system of a nation—cannot be properly understood by an *ad hoc* set of studies of the various elements that make up the system. The interrelationships between the components are so important and the variability created by the total environment in which the defined system operates so persuasive that always the whole is more complex than the sum of the individual parts. Within this concept it will be evident that the definition of the system under study is crucial (Anderson, 1974), particularly, as we have stressed throughout this book, the precise location of the system's boundary. Outside the boundary the 'environment' in which the system operates is both uncertain and difficult to predict while inside the boundary we have perceived a hierarchical structure of subsystems and sub-subsystems each with a similar autonomy to the major system under study. Thus the subsystems are considered to interact only through their conjugation in a higher system and we have seen in earlier chapters some of the implications of this concept on our modelling work.

The concept of systems theory has found expression through the techniques of modelling which we have explored previously in this text, though, as mentioned in Chapter 1, the model does not always have to be symbolic in form to be of value in describing a system. However, as we have seen, the computer-based simulation model has important advantages in this regard. Having now developed methods for the construction of such models and explored the appropriate experimental procedures in using them, we should now turn our minds to feasibility of application. Three broad functions can be discerned:

1. To provide an objective basis for assessing and assimilating available information about the system.

2. To direct research into important (relative to the complete understanding of the whole system) areas of the system about which current knowledge is sketchy or uncertain and to provide a platform whereby the results of this research can be immediately assessed (and applied). For some systems this job of probing into structure and functioning in an objective manner is sufficient: for others, perhaps of a more applied nature, a third function is essential.
3. To assist in the management control or development of the system. In agriculture this involves the application in extension and farmer decision making as well as in assisting policy direction and monitoring.

The first of these functions is really encompassed by the model-building process during which data are reviewed and assembled and requires no further explanation. The remaining functions are related to two major areas of application which will be presented in this chapter.

Systems Involvement in the Research Process

The deployment of research resources in agriculture is occupying decision makers in government in many countries (Fishel, 1971; Russell, 1973). The cost of agricultural research is a major item in national budgeting and the response to research effort is often less than satisfactory (Tollini, 1973). Part of the problem relates to the natural reluctance of the farming community to adopt new practices and technologies; the farm is not solely or mainly a business, nor is it the kind of organisation which is geared to take the risk of early adoption of research findings. Nevertheless, governments are looking for improvements in the efficiency of farming output while in the main relying on the traditional family structure of farming.

There will be a continued need for technical research but it is important that research resources are appropriately directed. Farmers on the whole are reluctant to be involved with research personnel, particularly but not exclusively, in less developed countries where significant social barriers exist (Andersen, 1974). In cases where collaboration occurs, the two parties are not always good bed-fellows; farmers tending to make judgements on their own or localised experience, researchers tending to be concerned with scientific development. Some countries fare better than others in this interaction but these tend to be the more agriculturally advanced.

Dillon (1976) in distinguishing between 'pure' and 'applied' research has stated that

> 'as soon as we look at a piece of agricultural research either in the context of the higher systems in which it is embedded or in terms of the lower systems which it influences we can see implications which go unrecognised in the traditional disciplinary approach to research.'

This basic premise of systems research has been largely unheeded by traditional researchers who have argued that their understanding of the discipline involved has been adequate alone to direct further studies. Dillon (1976), however, makes it clear that the systems approach can provide an indication of where research is needed: not only may significant data gaps be exposed, but also broader issues such as rural equality and the recognition of disadvantaged regions which are outside the remit of traditional research may be studied.

A CONCEPTUAL VIEW LINKING SYSTEMS MODELLING AND APPLIED RESEARCH

The systems researcher in agriculture has not been silent about the part to be played by his skills in integrating research and farming practice and in assisting with the direction of applied research. Two basic issues are obvious problems in directing current applied agricultural research.

1. Research programmes are established without necessarily undergoing a rigorously quantitative analysis of the farming to which they are supposed to be relevant. The deceptively simple task of identifying distinct farming types in a region or country so that research programmes can be orientated to meet specific requirements has not been developed. Yet it is patently obvious that unless clear knowledge of the productive structure of the farming industry is available and census information provides numbers of farms in each productive class, research direction will be, at least, haphazard (Chudleigh, 1976). Setting of priorities in research must be related to the benefits likely to be generated from alternative research programmes and the establishment of these benefits depends on estimating the effect of the new technology (research findings) on different farming types. It is axiomatic that a classification of farms according to productive capability is a first step in forging the link between systems researcher and the farmer.
2. The recommendations flowing from the completed experiments are

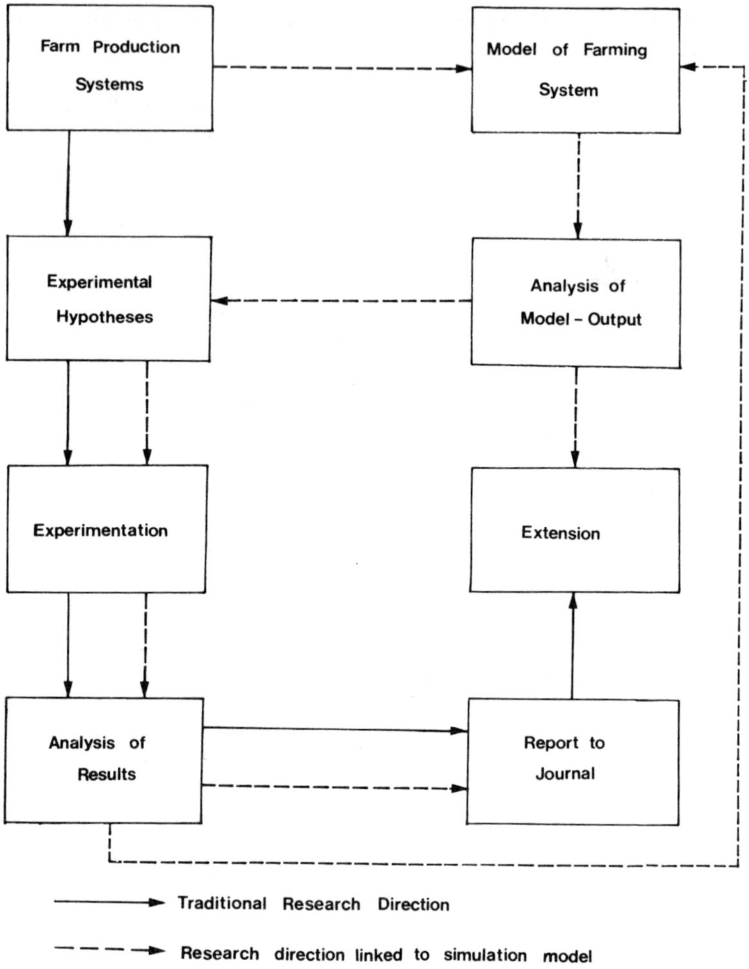

Fig. 7.1. Applied scientific research and systems modelling.

not presented in an assimilated form and in the context of the relevant farming methods. By and large, research findings are left for the farmer or more usually his adviser to evaluate and apply. These are complex and risky operations and, because they are left in the hands of the individual farmer, application of research findings is undoubtedly slowed down. The gap continues to widen between what is technically feasible (and perhaps economically desirable) and what is practised.

Conceptually both these problems are indicated and partially remedied within the simple framework shown in Fig. 7.1 in which systems modelling is linked with the traditional pattern of experimentation. The model in Fig. 7.1 is assumed to be a representation of a farming system or enterprise. The construction of the model as we have noticed is a procedure during which much is learnt about the 'gaps' in available information. Further information concerning areas in which there are data weaknesses will come to light in subsequent validation of, and experimentation (particularly sensitivity analysis) with, the model and in this way some attempt at research priorities may be established. Because potential research findings can be assessed within the model, priorities for research can be established relative to the likely rates of adoption of alternative findings, the objectives of farmers in the region, the forecasts of market development for alternative commodities, etc.

We will examine a theoretical exercise in sensitivity analysis. We might have two input parameters P_1 and P_2 which we wish to examine in terms of their effect on an output criterion. The feasible range for each of these parameters is divided into 100 parts and represented on the horizontal axis of a graph, as in Fig. 7.2. The vertical axis represents the value of the output criterion and the parameter whose changing level over the percentile range results in the greatest change in the output criterion would be placed at the top of the provisional research priority test.

In Fig. 7.2, parameter P_1 clearly causes a greater change in the value of the output criterion than does P_2 over their feasible ranges of operation. The position in the provisional listing can either be assessed graphically or by applying regression analysis to determine the relative slopes of the curves.

As an alternative approach, the barley-leaf-rust model was tested (Teng, Blackie & Close, 1977) for sensitivity to three parameters:

1. length of latent period;
2. length of infectious period;
3. rate of spore production.

In this case the output criterion was defined as percentage disease severity. The experiment used a factorial design on the three parameters with two levels of spore production, three levels of infectious period and three levels of latent period. Values for the variables were within the range known for cereal rusts. The simulated epidemic assumed optimum dew conditions for germination and the activities of liberation, survival and deposition were treated as stochastic events. Each epidemic was replicated three times. Table 7.1 provides the abbreviated analysis of variance results indicating that spore production rate and latent period were significant in affecting the simulated epidemics while infectious period did not have any significant influence. The length of latent period appears to be the dominant factor in determining the rate of build-up of rust-disease epidemics for the conditions specified.

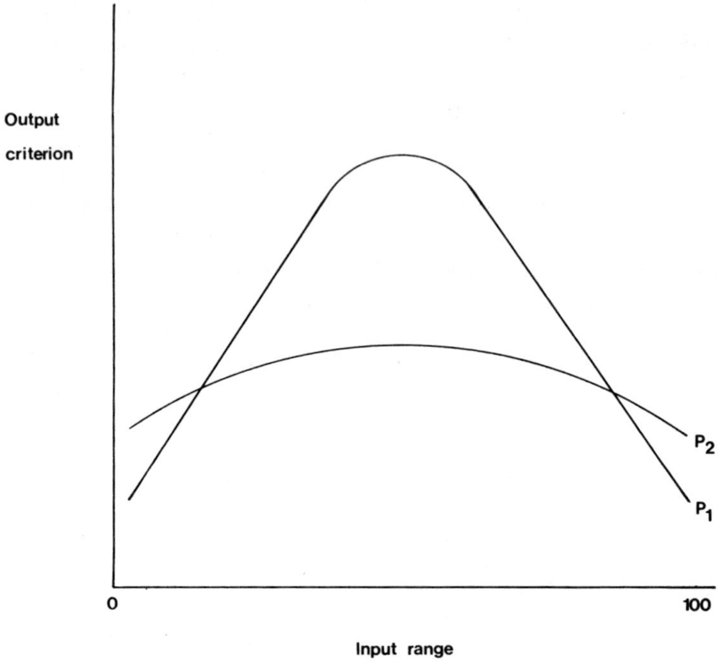

Fig. 7.2. Sensitivity results from varying two model-parameters.

The procedure outlined above becomes more complex when significant interaction between parameters occurs. Furthermore, it becomes difficult to interpret the results of a sensitivity analysis when a number of output criteria are involved. It is quite possible for a parameter to influence one output criterion and not another. The provisional testing then is usually not a simple ranking but, nevertheless, does provide an objective basis for examining the alternative research areas. Such a testing can only be

Table 7.1. Abbreviated analysis of variance of a simulated factorial experiment to determine sensitivity
(After Teng, Blackie & Close, 1977.)

Source of variation	*d.f.*	*Sum of squares*	*Mean square*	*F-value*
Latent period	2	8·834	4·417	605·88***
Infectious period	2	0·012	0·006	0·859
Rate of spore production	1	0·126	0·126	17·30***

provisional in character because other factors must be taken into account before research priorities can be finally established. The sensitivity analysis only gives an indication of the benefit likely from further experimental information about a given set of parameters. It does not consider the feasibility or costs of deriving these data or of existing equipment or skills available for research. Final decisions about research priorities must obviously be concerned with overall benefit from research carried out within the allocated research budget.

Finally, not only does sensitivity analysis help define research priorities but also the model-building process indicates the exact form in which the data can be most readily used. Therefore simulation can assist not only in the strategy of research planning but also in directing tactics. Results from experimentation would be disseminated by the traditional means, but also would be assimilated and assessed within the model, before complete recommendations could be made in the context of the farm type involved. Eventually, the model may be incorporated into a management-information system which can be used by extension workers and farmers to study the implications of given research findings in their individual farm situation. More will be said about this later in this chapter. The model then is carrying out two major tasks:

1. guiding the establishment of research priorities in the hope of getting the most satisfactory benefit–cost ratio for the resources being used. It should be appreciated that the terms of the benefit–cost ratio are unlikely to be solely technical and in most cases will be partly financial with strong social overtones;
2. acting as a medium whereby research findings may be accumulated, assessed and evaluated for individual farmers, so that the extension process may be improved.

The framework can be extended and classified by reference to Fig. 7.3 (Anderson & Dent, 1972) in which the right-hand segment outlines the conventional application of the scientific method in applied agricultural research and the left-hand segment presents the steps in systems research.

No matter which route (conventional scientific research or systems research), facts must be assembled regarding the problem at hand. This stage is represented by block 1 and typically involves a blend of induction, deduction, inspiration and informal appraisal. The links 9–1 and 1–14 represent the potentially useful route by which systems simulation results can influence initiation of research on a real system. However, the role played by blocks 10 and 11 may be valuable to practitioners and directors of

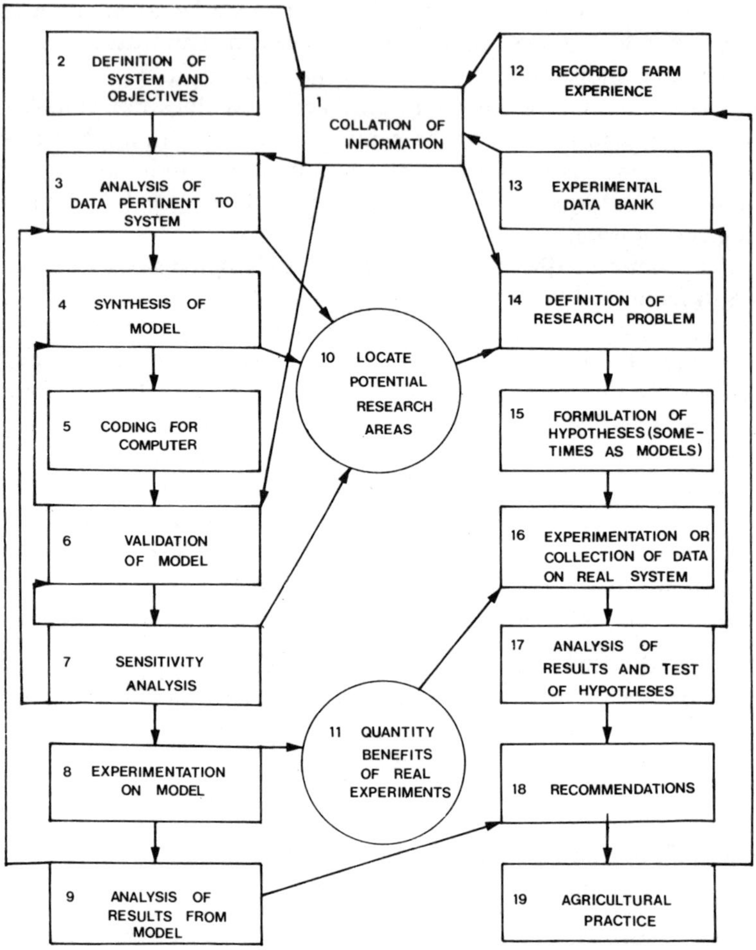

Fig. 7.3. A framework for linking applied research and systems modelling.

research. The location of potential areas of research will probably be proved through the links 3–10 and 4–10. Here the almost universally encountered gaps in knowledge that are revealed by system analysis and simulation will be brought to light. This systematic organisation of knowledge about the parts of a system is usually the most revealing stage in locating shortcomings in data and theory to the 3–10–14 linkage.

This concept ideally extends the boundary of applied research in agriculture beyond the purely technical. Thus, in order to be of maximum value, the modelling effort should have access to disciplines such as extension, rural sociology, marketing and trade economics, for these create the environment in which farming operates. Unfortunately, this kind of integration currently is not well developed. Those models that have been developed either for aiding research direction in some way or for some general farming extension purpose have all been conceived within the broad agricultural discipline and generally in the universities or research institutes. The implication is that there is sufficient 'awareness' of the farming scene to permit the modelling to proceed. This 'awareness' tends to be lacking in objective assessment of

1. the population of farmers to whom the work applies and whether they can be grouped in any way according to production system;
2. the present real problems on these farms;
3. the present production levels attained;
4. the existing range of aptitudes, skills and management capability;
5. the attitude of the farmers toward change.

In the almost universal absence of these data, any models built to link with the traditional research process can at best be broad representations for a specified type of farming in a specified area. Again it is a question of the incisive placing of the system boundary which determines the real value of the modelling work. The incorporation of the above listed socioeconomic factors will extend the assistance which a model can provide in the direction of research priorities and hence in the efficient deployment of research resources.

SYSTEMS MODELLING AND THE DIRECTION OF FUNDAMENTAL RESEARCH

There is no doubt that fundamental research is a costly operation and that results cannot easily be programmed. However, there is a body of opinion which argues that fundamental research, as much applied research, requires objective guidance in establishing programmes for maximum effectiveness. Recent evidence suggests that guidance may be provided by systems concepts. The framework in which this guidance may be provided is similar in form to that shown in Fig. 7.1, except that the involvement will be directed towards a biological system and that the objective will be a better understanding of the system so that it can be more effectively understood and subsequently controlled. Figure 7.4 sets out the schematic relationship. Perhaps the most important practical point to note is that the model-structure will usually be quite different to that for the models in Fig. 7.1. The model-types in Fig. 7.1 are usually directed towards the prediction of major production parameters, there being little attempt to explain the mode of operation of cellular or biochemical processes. For the models of Fig. 7.4 an entirely different approach is required in which the model is explanatory in nature so that theoretical statements concerning the processes represented are included (Baldwin & Smith, 1970).

This distinction in model-type has a number of implications in all stages of model-development from the initial formulation of concepts, through selection of computer language, to methods of validation. Increasingly, fundamental biological scientists in agriculture are finding that their research programme can be more pertinently directed if it is integrated with an explanatory type of simulation model: precise data deficiencies and hence experimental approach are suggested and a clear understanding of the whole system frequently projects a line of investigation not previously appreciated. Three briefly stated examples emphasise the point:

1. In attempts to simulate growth and development of ruminants, a main component is the representation of the regulatory mechanisms in adipose tissue. Until modelling work was attempted (Baldwin & Smith, 1971) the complete lack of data regarding triglyceride turnover rates in ruminants was not recognised, nor were their importance in understanding the overall functioning of the animal appreciated. Analysis of the simulation modelling has prompted research to move in an (as yet unsuccessful) attempt to rectify the deficiency.
2. Perhaps one of the most important areas restricting the understanding and development of grazing systems is that of food intake. The concepts are still not finalised but the modelling work of Mertens (1973) on the kinetics of cell-wall digestion opened up the field of research and directed investigations into rates of food passage and food particle size as important determinants of food intake.
3. It is worth mentioning that predictive models may also be constructed at a level which potentially permits fundamental research guidance in the manner expressed above. However, without an attempt to represent the components in the model in a mechanistic manner there are obvious limitations for this type of model. The work of Bywater (1976) concerned with developing a whole-animal simulation for dairy cows may be cited as an example. As a result of this project it is clear that the understanding of endogenous glucose synthesis during early lactation (when the animal is using feed reserves to supplement ingested food) is totally inadequate. This period of lactation is crucial to efficient milk production and yet it transpires that the basic nutritional components are not even conceptualised.

Figure 7.4 has additional components over and above the general layout of Fig. 7.1 and this relates to the use of explanatory or detailed predictive models as a base for establishing input–output data which are important in farm-level or enterprise-level studies. Throughout the world, a great deal of cost and effort is spent on applied research either in 'controlled' and

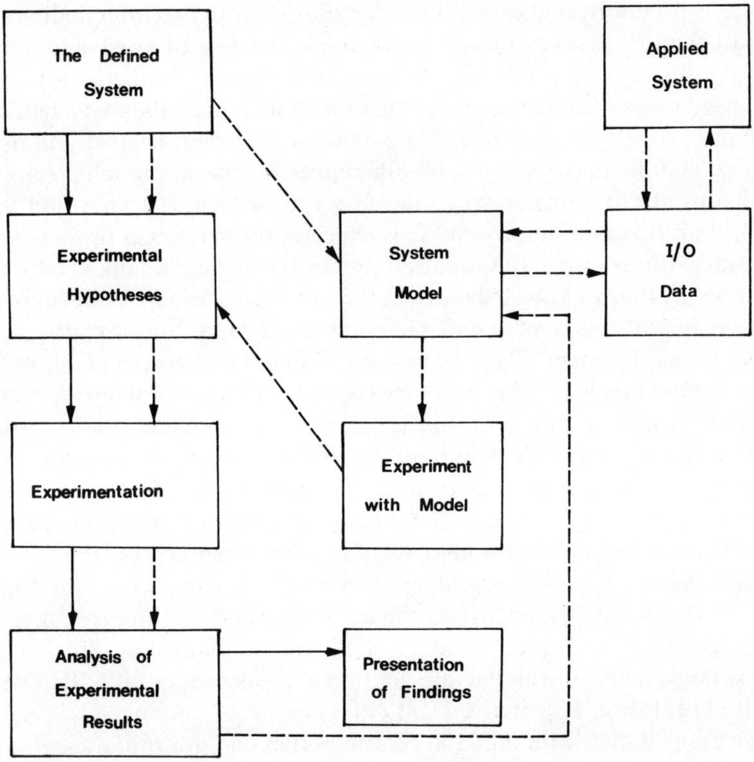

Fig. 7.4. Fundamental scientific research and systems modelling.

partially 'controlled' experiments or in field trials to generate these input–output data. For much crop and grazing livestock research (at least), experimental difficulties in generating meaningful data and interpreting them for farming practice are conveniently suppressed. This is understandable because the agricultural industry has demanded applied technical information and no other way of meeting this requirement beside carrying out applied experiments has been available. Some of the problems are

1. the variation of results from field experimentation between seasons and between regions;
2. the time required to complete a trial even in a restricted number of seasons;
3. the cost in time, resources and effort in conducting such experiments;
4. the interseasonal interactions involved;
5. the deficiencies of measuring techniques.

There is no doubt that possibilities for shortcutting this often tedious and unsatisfactory process now exist through the use of tried and proven simulation models of either an explanatory or detailed predictive type. Care will need to be exercised to ensure that the models are validated adequately and are properly maintained. The potential, however, is great and in due course should make significant differences to the approach to applied agricultural experiments: speed, accuracy, economic efficiency and scope should all be vastly improved. This expectation may seem optimistic but consider, for example, the situation presently existing in milk production: the present state of knowledge about the response of different breeds of cow to feed quantity and type at different points during lactation is inadequate for effective management. Even the first question to be answered in this regard, that of how much of different feed types will the cow actually consume if offered, would be an enormously costly and time-consuming task to complete experimentally. Consider the present need to evaluate the use of crop by-products, new cultivars and alternative types of forage crops and various conserved products in ruminant production in attempts to improve the efficiency and reduce the land use of this type of enterprise: the enormity of the task is beyond present applied research capability, not from a technical point of view but in relation to the time and resources required. An appropriately structured model based on fundamental biochemical, physiological and metabolic information such as proposed by Baldwin & Smith (1971) and Bywater & Dent (1976) could be employed to generate these applied data with minimal resources (beyond appropriate computer facilities) and in a comprehensive way so that not only feed input, animal liveweight changes and milk output is known but also the metabolic pathways involved could be monitored.

While such developments are perhaps a little way in the future it should be understood that movement toward appropriate models is progressing in a number of areas and the potential should be starting to influence decisions made now about applied research resource deployment.

Systems Involvement in Farmer Decision Making and Business Control

For many farmers and particularly those in developed economies, in any assessment of research findings, the farm manager has to determine how the new development will fit into his particular farm system and what influence it will have on his total profit and cash flow. The manager's concern for the overall management and control of the farm business will, from time to time, involve the assessment and introduction of new technology. Systems research in the form of 'management-information systems' recently has been extending into the farm management and control area from its original base in industrial management.

A management-information system for farming is an integrated feedback mechanism which can be used

1. to develop plans for the farm that are technically feasible and economically desirable, and to assist with implementation of a chosen plan;
2. to analyse the results of previous action from farm records;
3. to examine these results in the light of the expectations from the plan;
4. to suggest adaptive procedures where results and plan deviate.

The information system can, therefore, be regarded as consisting of a 'planning package' and an 'analysis package'. The planning package is used to produce plans to generate expected outcomes from a selected plan and to investigate adaptive procedures when necessary. The analysis package processes current data on the progress of the plan and draws the manager's attention to unexpected deviations from the plan. The structure of such an information system aimed at the individual farm is depicted in Fig. 7.5 and the steps in its functions are, briefly, as follows:

1. Using the planning package examine alternative strategies for the farm including assessment of any technical innovation. The information provided by the planning package would include:

   (a) general economies of the changes;
   (b) the resultant cash flows;
   (c) the manpower/work load for the farm;
   (d) the level of indebtedness;
   (e) the technological efficiency, and other relevant measures to aid in assessment.

   In using such a planning package, the farmer is interested not only in the overall effect of any technical change, but also in the effects this change will have on the routine operation of his business during the 'phase-in' period. There is a real need for detailed technical and financial information about that crucial period of change and a properly constructed information system can fill this gap. There is a great deal of learning in adjusting to a new method of production

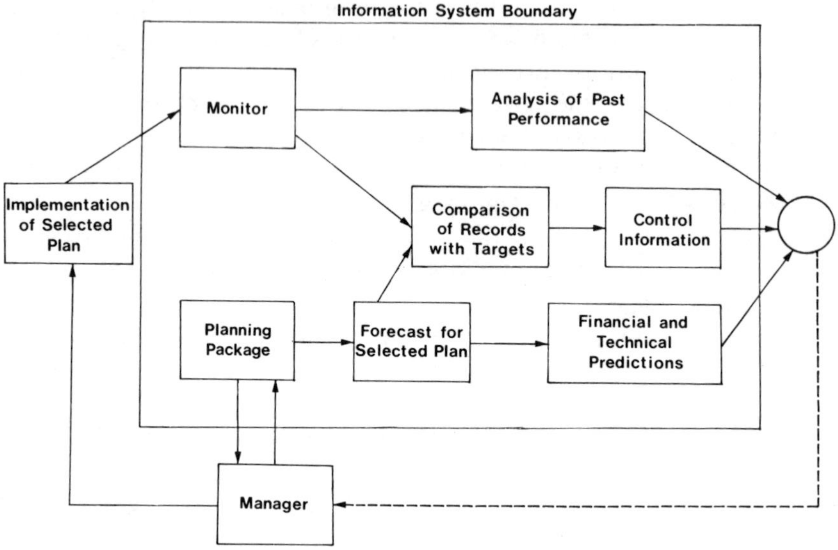

Fig. 7.5. The structure of an information system.

no matter how small a change is involved, and it makes sense to learn via an information system where possible.

2. On the basis of all information to hand, decide whether or not to adopt the change that has been investigated.
3. Whatever the decision in 2, the farmer will continue recording appropriate information at regular time intervals:

   (a) to keep the data on his unit held in the information system current;
   (b) to provide himself and possibly his consultant with an up-to-date analysis of actual performance.

   The time interval for recording will clearly depend on the type of farm (or farm enterprise) and might vary from a week or two to several months; the essence of success is the speed at which farm records are analysed and this *must* be enhanced by the information system.
4. Whether the decision made was to adopt a new strategy or to continue without change, the information system should then cause the planning package to produce production targets. Such targets are values for key elements in the plan which must be achieved if the

plan is to succeed. Obviously a consideration of time must be included in any relevant target so that the target is not simply a physical value but a value which must be achieved at some specified time in the future—for example, quantities of silage to be fed to beef fattening stock expressed in kilograms dry matter per head per month. Unless these targets are met over time, the farm will not be behaving as planned. Such targets are particularly important when a new farming programme has been introduced for which traditional targets are no longer relevant.

5. The comparison of targets with actual recorded (and analysed) figures form the basis for management control during the tricky period between adoption and absorption of a new innovation. The actual performance achieved in the farming operation is quickly determined and this compared with the appropriate time-based targets. Confidence may or may not be improved. But at least the farmer will know immediately about any adverse deviations from target; adaptive action can then be taken.

The general discussion of information systems at farm level can be pursued into the context of the barley-leaf-rust model and an information system for disease control on a farm. Figure 7.6 sets out the structure of the system in a similar format to Fig. 7.5. Now the alternative strategies to be explored will relate to different rotational options on the farm, the total area devoted to various crops, sowing dates for crops, irrigation water allocation, fertiliser treatment on the various crops, etc., and the assessment of the strategies would be in terms of yields of crops from each paddock, crop-loss estimates, cash flows over a specified period of simulation time. The planning package thus has to be a model which is capable of forecasting the assessment parameters not only in response to various alternative management strategies but also in relation to a changing climatic environment.

The information system now requires that crops on farms using the system are regularly monitored so that the forecast of crop yield and disease build-up can be up-dated during the crop growing season. By way of the information system a farmer could then see the likely effects of crop-protection procedures at various times on crop loss and profit: these can then be placed alongside predictions of the same parameters if no protection measures were instituted. Regular comparisons give firm guidelines for crop-protection action based on the benefit–cost ratio for action at any time.

Recall that one of the essential parts of a management-information system is a planning package which must be capable of reflecting the timing of management decisions. It must be able to incorporate the effects of new technology and to mimic in detail the response that would be expected from

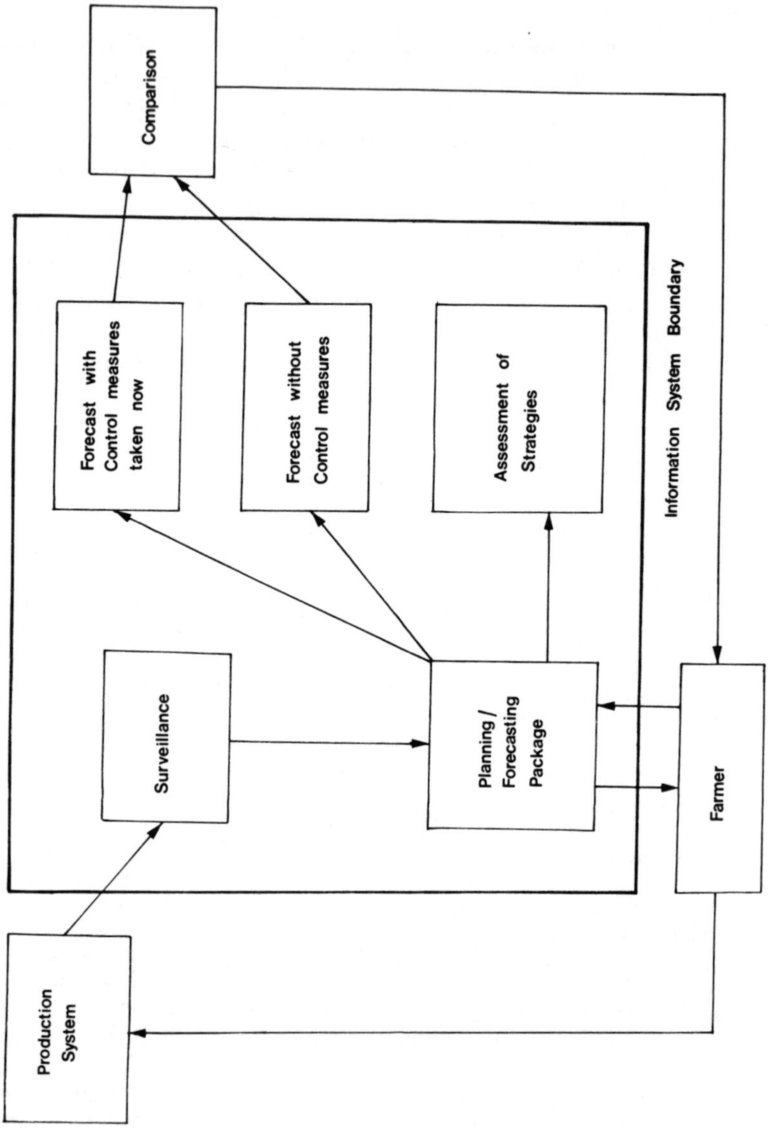

Fig. 7.6. *The structure of a crop-protection information system.*

the real farming system in relation to the whole range of management decisions. As it would be unthinkable, in terms of expense and organisation, to have a separate information system for each farm, the package must be applicable to a wide variety of farms or farm enterprises of a given type.

In essence, the planning package has to be a model which is capable of adequate representation of the appropriate farm or enterprise system which implies that it is dynamic and stochastic in form. As we have seen, computer-based simulation models match up to these requirements very well. Ideally then, the planning package should be a simulation model within the information system, but the prospect of building a computer simulation model for every farm in the information system is totally impractical. This difficulty has been overcome by the use of 'skeleton' simulation models. The 'skeleton' model—which is discussed in more detail below—has a low charge for development to the individual since it is applicable to a large number of farms and, therefore, the development costs can be shared between a large number of users. As we have seen earlier, it is the development phase of simulation modelling that incurs the major expense. The skeleton model relies on the fact that certain parameters form the basic logic of a given real farming system and that this logic is unchangeable from farm to farm. It can be thought of as a model of the lowest common denominator for all farms that might be associated with the information system (Blackie & Dent, 1973). Figure 7.7 represents the diagrammatic form of a skeleton model for dairy production. In fact any dairy enterprise can be directly associated with the skeleton in Fig. 7.7. Of course, a great deal of specific information is required to complete the picture for a specified dairy unit but these particular data can be 'plugged' into the general model of Fig. 7.7. Provided the skeleton takes care of basic biological processes, constants and essential logic, when it is linked with appropriated information (records) from an individual farm the resultant complete model will be tailor-made to that individual farm and will provide a comprehensive simulation of it. As any one farm uses the information system, its own background, planning and latest records are linked to the general skeleton model and this completed model is then an individual model of that farm.

Before proceeding it is worth noting that the skeleton-model concept has been applied in modular form to whole-farm simulation studies. Independent skeleton modules of farm enterprises are called as appropriate for the individual farms to form the skeleton core of a total farm model. An example of this approach is provided in Ritchie, Dent & Blackie (1978).

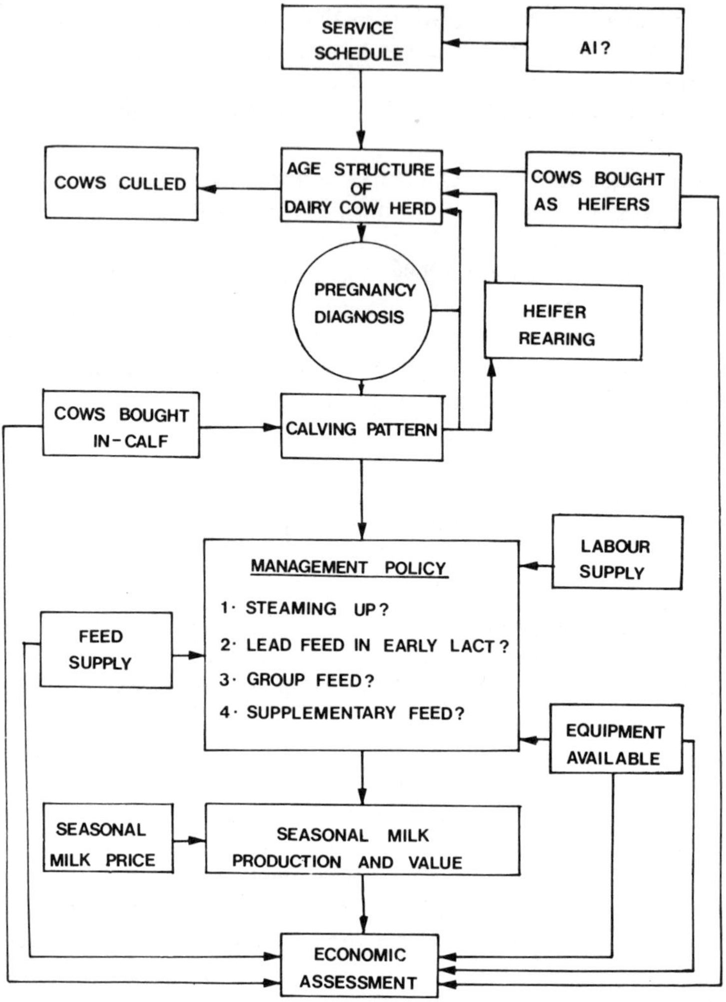

Fig. 7.7. A skeleton representation of milk-production enterprises.

Design Pointers for Successful Agricultural-information System Designs

The design and implementation of information systems is a subject in its own right and we can only briefly consider some of the salient points of system design in this text. One of the most impressive and successful agricultural-information systems is the CANFARM system first mooted by the Canadian Government in the late 1960s but not successfully implemented until 1975 (Thompson, 1976). Thompson describes many of the problems overcome during the design and early implementation stages, and other workers in this area, including the authors, have met similar problems. The main difficulties in implementing agricultural-information systems are

1. lack of user acceptance;
2. misinterpretation of data requirements either in the field or in the computer centre;
3. delays caused by the late arrival of field data, inadequate data preparation facilities at the computer centre or software 'bugs'.

In the main, the causes of these problems can be attributed to

1. the range of activities for which an agricultural-information system must cater;
2. communication difficulties between the parties concerned in the development and use of the system;
3. lack of experience in the use of computer techniques in agriculture.

As the CANFARM experience has shown, these difficulties, although severe, are not insuperable. We recommend the following practices in the design and implementation of an information system. They are a distillation of experience from developing information systems with a tight budget and with a small staff—a situation likely to be common in agriculture.

1. Pay particular attention to data-file design. The heart of any information system is its data files. If these files are damaged or lost, the information system loses much of its value and most of its credibility. The files should be designed so that any mistakes on the part of the user are caught before any permanent modifications to the files are made. File-management techniques are so interlinked with the particular computer being used that it is difficult to give instructions on how this may best be achieved. However, some points may be of assistance.

   (a) If code numbers are used, codes should not be similar. For example, a three figure code of 201 could be mistaken for code 207. Techniques exist for generating a series of code numbers which are unlikely to be confused (see Sprowls, 1976, for example).

(b) Hold on the file information concerning such items as the date of the last update, the user who last accessed the file and so on. It is then a simple matter to ensure that the input to any run contains some or all of this information. If the two sets do not match, something is wrong and the run terminates after printing out the reason for the error. This prevents a file being updated twice or an update missed due to an operating-system fault.

(c) Have a security code on file to prevent unauthorised or unintentional modification of the files by other users. A simple password which is periodically changed may be suitable for most cases.

(d) Include in the information-system design a report which is produced by the system for each run. This report will inform the user as to any major or unexpected file alterations, input errors or data-validation failures.

(e) Design the files so that if errors do get into the files—and they will—the files can be amended easily.

2. Start with a simple design. The evolutionary approach to information-system design in agriculture is the simplest and least traumatic in our experience. It is extremely difficult to foresee the costs and problems of system development. A simple, well documented and reliable system provides an active base for development. Improvements can be proposed, tested and then implemented while the system still provides a continuing information service to the farmer. Communication is improved between system designers and the users and the continuing improvements stimulate interest in the system.
3. Allow plenty of time for discussion. Many farm managers and extension workers are totally unfamiliar with computer technology. Hence they do not appreciate the problems associated with even minor changes to an information system or with the need for total consistency in terminology. This leaves plenty of opportunity for misunderstandings between the system designers and the users. The designers of the system should become as familiar as possible with the operating environment of both farmers and extension workers.

4. Concentrate on input and output form design. If the user does not understand the data he is required to collect or the information returned is not easily comprehensible, then the system is unlikely to succeed. Frequently, the system designers will be forced into several modifications of input and output format—a tedious and often frustrating process, particularly in non-business orientated languages such as FORTRAN. However, due to the difficulties of establishing an effective feedback between the farm environment and the computer centre, we have found it necessary to design a form, test it in practice and then redesign it in the light of our experience and users' comments. It has proved difficult to get farmers and advisory workers to examine forms adequately until they actually use them.

Future Developments in Agricultural-information Systems

Dunn (1974) and Eisgruber (1973) have pointed out the deficiencies in modern data-collection methods. Similar reservations have been expressed by the American Agricultural Economics Committee in Economic Statistics and by Bonnen (1975) in his presidential address to the American Agricultural Economics Association. While these problems are beyond the scope of this text, the authors cited above make a strong case for a new look at data-collection methods in agriculture and their recommendation is that information systems can and should be constructed to meet not only the requirements of the individual farmer but also the information needs of policy makers, statisticians, consumers and industry leaders. To quote from Bonnen's address:

> 'We face a major problem in the redesign of agricultural-information systems and share in a great opportunity again to contribute to agriculture and the social sciences in a fundamental way, much as agricultural economists did in the early days of econometrics and, in the late 1920s through 1940, when they developed major information systems to manage and ameliorate the problems of a Depression and a World War.'

We consider the type of farm-information system proposed in this chapter can go some way towards providing industrial and national information. The operation of such a management-information system for single enterprises or at whole-farm level involves data from every unit involved being sent at regular intervals to the computer centre for linking with the skeleton model. Data concerning yields of crops, livestock performance, area of land under different crops, numbers of different types of livestock, quantities of different types of inputs are involved as well as, in some cases, financial information about input costs and selling prices: management practices are also detailed. The computer files therefore hold current information in the form of both raw and analysed farm records. Since these data refer to individual farms, an important part of the computer organisation must be concerned with the confidentiality of information held. Clearly, individual farm data must be confidential to the farmer and his adviser, otherwise the integrity of the scheme would be in jeopardy (Dunn, 1974). However, aggregate data referring to similar enterprises or to farm types can easily be assembled and analysed within the computer installation without loss of individual confidentiality.

Data retrieval from an industry as widely dispersed as agriculture is a major problem for policy makers in government, commerce and industry serving agriculture. The demand for information to permit rational policy and marketing decision making by these agencies is growing in all countries of the world and the farm-based management-information system potentially provides a unique and detailed data bank. The CANFARM system in Canada is currently concentrating its efforts on its data-bank facilities which represent the most recent innovation to the system. Thompson (1976) reports that requests for data from the data-bank facility range from price analysis to the usage of electricity on farms in Quebec. Considerable potential exists for the expansion of this type of service.

The attraction of this concept is that it reduces the need for statistical survey (although obviously it does not eliminate such work). Because the farm manager has a direct return from his recording efforts (in terms of valuable on-farm information for business control) his records are likely to be accurate. The data then available for policy decisions are current and reflect the existing environment of the industry. The danger exists that only the better farmers (or possibly the larger business) will use such services thus giving a biased sample for analysis. Alternatively farmers having access to the information system, because they have a potent management aid become, over a period of time, a unique sample from the population whose statistics are no longer representative. However, such problems are not insuperable.

Examples of existing and planned information systems will provide some background.

1. A commercial feed manufacturing company operates information

systems for pig producers and for dairy farmers. While the management aspects of the system have proved of value to farmers, the company itself has derived benefit

(a) because its advisory personnel are supplied with current management data for all farms visited and therefore have improved their efficiency: the cost effectiveness of the advisory service to the company has been improved;
(b) aggregate statistics about the performance of stock throughout the country on different brands of feed are directly available, thus giving the company information on the nutritional input–output relationships achieved by farmers, the economic relativity of their various brands of feed and the general financial viability of different types of farms.

2. A farmers' marketing organisation is planning an information system

(a) because it wishes to improve the efficiency of its extension officers;
(b) because it requires cost of production data for its product by region and by farm size in order to assist the statutory fixing of minimum product prices and the most effective way of gaining this is via an information system.

3. A government agency established an information system for dairy farmers once again to assist its advisory officers but also because it required aggregate data for its marketing board price-fixing procedures and for its policy makers in discussing the structure and state of their dairy industry.

Skeleton Models in Applied Research Direction

The type of model inserted within the framework of Fig. 7.1 could well be of a skeleton type. As previously mentioned, an important feature of a skeleton model is that it can be so organised to represent any manifestation of an enterprise or farm. The potential results from a projected research proposal could then be assessed within the skeleton model while set-up to represent a series of different farm types to which the findings could be relevant. From this analysis, the research director should find assistance in estimating the value of the research programme. This approach can be illustrated with regard to the pig-production model and by reference to a number of nutritional experiments.

The work of Perez-Aleman, Dempster, English & Topps (1971) indicated that for every 10 per cent addition of dried poultry manure in the diet of the growing pig, average daily gain was reduced by 0·02 kg and feed conversion efficiency by 0·25. The question to be answered, therefore, was whether, on a typical pig-production unit, it would prove more profitable to feed a cheaper diet containing dried poultry manure and accept a poor fattening-pig performance or to feed conventional meal. This is a complex decision since a slower growth of fattening pigs means that the pigs will occupy the fattening pens for a longer period. The farmer has three alternatives open to him, each of which has different implications on the profitability of the unit:

1. expand his fattening facilities so as to be able to accommodate more fattening pigs at any given time;
2. reduce the size of the breeding herd so as to decrease the flow of pigs into the fattening pens;
3. accept overcrowding in the fattening pens and an even greater deterioration in the performance of fattening pigs.

Simulation experiments examined various inclusion levels of dried poultry manure with alternative management strategies and various cost–price ratios. The results from this set of investigations showed that at above 10 per cent inclusion of dried poultry manure in the diet, performance was so severely affected that the practice was unlikely ever to prove profitable. Under some extreme cost–price situations, low inclusion rates could prove profitable. Had this simulation investigation been completed prior to the experiment being carried out, undoubtedly changes would have been contemplated, at least to the experimental treatments.

But it is conceivable to use the skeleton model one step earlier than this, in a situation whereby the research director can 'experiment' with the model-representations of the various farm types with which his research station is concerned to explore the sensitivities of these farm systems (in profit or output terms) to the findings of *possible* research programmes.

Refer again to the discussion surrounding Table 7.1. The results, it will be recalled, indicated that by far the most significant factor in determining the build-up of epidemic was the length of the latent period. Since this period is the result of host–parasite interaction, a recommendation from this sensitivity exploration was that breeding for rust resistance in cereals could be most effective if breeders concentrate on developing varieties which cause a long latent period to result.

The flexibility of the skeleton-model concept is of major significance since it obviates the need to construct a series of specialist simulation models. Provided an information system is operating, the research director does not need any additional facility than that needed for management purposes by the farmer.

Final Synthesis

The essence of this chapter has been to examine areas of impact of systems concepts in agriculture and to view the use of modelling within these areas. Attention has been focused on the direction of research and the potential for effective deployment of research resources and on the use of information systems to assist management at farm level, the extension services and policy makers. Figure 7.8 attempts to summarise the discussion and to view the main elements as subsystems in the agricultural research, development and production system.

Consideration of Fig. 7.8 is perhaps best initiated at the information systems and the skeleton simulation models required to drive them. As well as recorded farm information, the skeleton models require applied biological relationships as basic data within their structure. Where, as often will be the case, relationships of an appropriate form, developed in an acceptable way, under suitable conditions are not directly available from

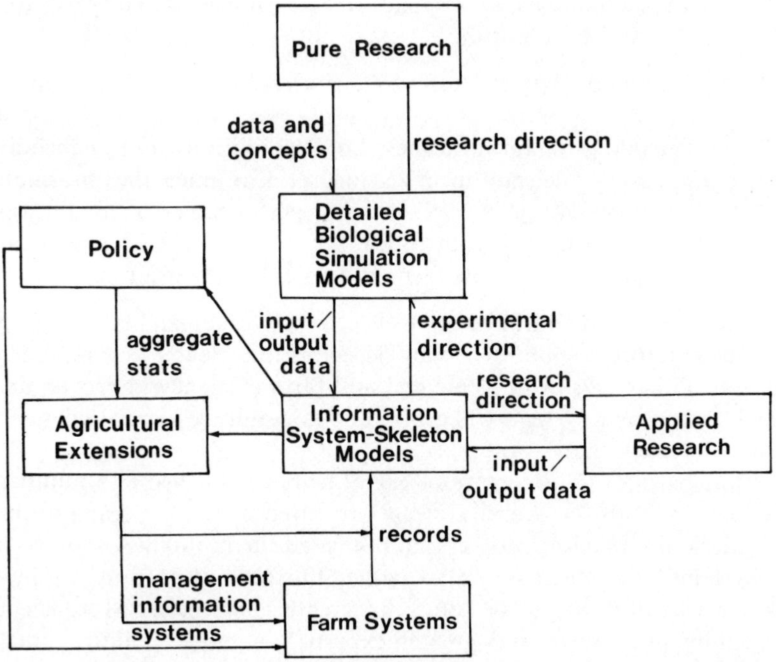

Fig. 7.8. Information systems, simulation modelling, agricultural research and farm practice.

previous applied research, new data may be developed either by traditional applied research or by reference to a detailed biological model. In either case, the needs of the skeleton model will direct the process as suggested in Fig. 7.1. Research direction via sensitivity analysis and farm-production system classification is provided by the skeleton model while data are accepted by it from applied research or from detailed biological models. In turn, assistance in assessing research priorities in fundamental research is provided by the detailed biological model which uses concepts and data from the on-going research (Fig. 7.4).

The role of information systems in gathering production statistics about farming or farming systems has been outlined in this chapter, and if many farmers are involved in an information system, this surveillance leads to the possibility of gathering aggregate current production statistics on a scale not previously possible and on a time scale hitherto unachievable. The availability of such aggregate data has immense potential from a farm policy point of view either by government or by commercial firms. In either case, agricultural extension agencies are bound to be influenced. But it would be hoped that the major impact on extension would be by way of the information system permitting

1. planning of farm strategies to meet the objectives of the individual farmer;
2. controlling the farm business both biologically and financially;
3. improved efficiency in extension services since the information systems would provide a record of performance of the farm in a current form. This provides the extension officer with firm information about the farm and its efficiency prior to his visit.

Farming systems then will be more effectively planned and controlled and should therefore respond to policy decisions in a clearer and more rapid fashion. Policy can be implemented and farm efficiency improved under which forces farm systems will evolve and hopefully resources will be used more effectively than at present.

While parts of the structure of Fig. 7.8 are established in a number of disciplinary fields the overall concept presented is merely a framework for consideration. Exciting progress has been made in the linking of some of the subsystems and much is yet to come. Further integration within the framework will lead, in the future, to greater efficiencies. More effective modelling procedures and greater expertise in model-building are key elements in the structure. We hope this book can make a contribution to the vision. Models built with an appreciation of both their potential and their relationship to the agricultural industry as a whole will, we believe, be of a standard where they can be used in the applications foreseen in this final chapter. This book is intended to be one step towards the achievement of such performance in agricultural modelling.

References

Andersen, P. P. (1974). Towards a workable management tool for resource allocation in applied agricultural research in developing countries. Mimeo, C.I.A.T., Calif., Columbia.

Anderson, J. R. (1974). Simulation: methodology and application in agricultural economics. *Rev. Mktng Agr. Econ.*, **42,** 3–55.

Anderson, J. R. & Dent, J. B. (1972). Systems, simulation and agricultural research. *J. Aust. Inst. Agr. Sci.*, **38,** 264–96.

Baldwin, R. L. & Smith, N. E. (1970). Application of a simulation modelling technique in analysis of dynamic aspects of animal energetics. *Federation Proc.*, **30,** 1459–65.

Baldwin, R. L. & Smith, N. E. (1971). Intermediary aspects and tissue interactions of ruminant fat metabolism. *J. Dairy Sci.*, **54,** 583–95.

Blackie, M. J. & Dent, J. B. (1973). A planning and control system for the small firm: a development of operations research in agriculture. *Z. Operations Res.*, **177,** 173–82.

Bonnen, J. T. (1975). Improving information on agriculture and rural life. *Am. J. Agr. Econ.*, **57,** 753–63.

Bywater, A. C. (1976). Simulation of the intake and partition of nutrients by the dairy cow. II. The yield and composition of milk. *Agr. Systems*, **1,** 261–80.

Bywater, A. C. & Dent, J. B. (1976). Simulation of the intake and partition of nutrients by the dairy cow. I. Management control in the dairy enterprise; philosophy and general model construction. *Agr. Systems*, **1,** 245–60.

Chudleigh, P. D. (1976). The use of classification and description of animal production systems in the formation of priorities for agricultural research in Kenya. *Agr. Systems*, **1,** 281–300.

Dillon, J. L. (1976). The economics of systems research. *Agr. Systems*, **1,** 5–22.

Dunn, E. S. (1974). *Social Information Processing and Statistical Systems—Change and Reform.* New York: Wiley.

Eisgruber, L. M. (1973). Managerial information aid decision systems in the USA: historical developments, current status and major issues. *Am. J. Agr. Econ.*, **55,** 930–7.

Fishel, W. L. (1971). *Resource Allocation in Agricultural Research.* Minneapolis: University Minnesota Press.

Mertens, D. R. (1973). Application of theoretical mathematical models to cell wall digestion and forage intake in ruminants. Ph.D. Thesis, Cornell University.

Perez-Aleman, S., Dempster, D. G., English, P. R. & Topps, J. H. (1971). A note on dried poultry manure in the diet of growing pigs. *Animal Prod.*, **13,** 361–4.

Ritchie, I. J., Dent, J. B. & Blackie, M. J. (1978). Irrigation management: an information system approach. *Agr. Systems*, **3,** 67–74.
Russell, D. G. (1973). *Resource Allocation System for Agricultural Research.* Sterling: University Sterling Press.
Sprowls, A. C. (1976). *Management Data Bases.* Santa Barbara: Wiley/Hamilton.
Teng, P. S., Blackie, M. J. & Close, R. C. (1977). A simulation analysis of crop yield loss due to rust disease. *Agr. Systems*, **2,** 189–98.
Thompson, S. C. (1976). CANFARM—A farm management information system. *J. Agr. Admin.*, **3,** 181–92.
Tollini, H. (1973). Planning agricultural research: concepts and practice. *Proc. 15th Intern. Conf. Agr. Econ.*, 296–305.

Index